Reviews of Plasma Physics

Reviews of

Plasma Physics

Edited by V. D. Shafranov

Volume **23**

© KLUWER ACADEMIC / CONSULTANTS BUREAU
b NEW YORK · BOSTON · DORDRECHT · LONDON · MOSCOW

Camera-ready copy prepared in Russia by:
K.A. Postnov and B.M. Smirnov (translators)
M.J. Dudziak (language editor)
M.S. Aksent'eva (managing editor)
E.V. Zakharova (bibliography editor)
and **Vsevolod Kordonskii** (desk editor) of
Uspekhi Fizicheskikh Nauk

ISSN 0080-2050
ISBN 0-306-11069-5

© 2003 Kluwer Academic / Plenum Publishers, New York
Kluwer Academic / Consultants Bureau
233 Spring Street, New York, N.Y. 10013

http://www.wkap.nl/

10 9 8 7 6 5 4 3 2 1

A C.I.P. record for this book is available from the Library of Congress.

Printed in the United States of America

CONTENTS

PLASMA MODELS OF ATOM AND RADIATIVE – COLLISIONAL PROCESSES

V. A. Astapenko, L. A. Bureyeva, V. S. Lisitsa

1. Introduction 1
2. Plasma characteristics of atoms in phenomenological and kinetic models of atoms. Dynamic polarizability of atomic structures 8
   2.1. Brandt – Lundqvist local plasma frequency model 9
   2.2. The response functions and the static polarizability of a Thomas – Fermi atom 15
   2.3. Kinetic model of the dynamic polarizability of atoms 18
   2.4. Quantum calculations of the dynamic polarizability 20
3. Static and polarization radiative channels in collision of charged particles with atoms and plasmas 24
   3.1. General consideration for bremsstrahlung of fast particles on atom 24
   3.2. The dynamic form factor formalism in description of radiation from fast particles in a plasma 37
4. Plasma models for photoionization of atoms and ions . 47
   4.1. Application of dynamic polarizability models to calculations of the photoeffect on complex atoms 47

4.2.   Approximate quantum methods
         for photoabsorption cross section calculations .   51

5.   Bremsstrahlung and photorecombination of moderate
     energy electrons on multielectron atoms and ions . . .   57

4.2.   Radiative losses of electrons in scattering
         on neutral atoms   58

5.2.   Core polarization effects in emission
         and recombination of electrons
         on multielectron atoms   78

5.3.   The transition bremsstrahlung of thermal
         electrons on plasma ions   91

5.4.   Quantum calculation (by the incident particle
         motion) of the effective radiation
         on multielectron ions   97

6.   Polarization channels of fast particle radiation
     on atoms, in plasma, and in a dense medium 103

6.1.   Polarization bremsstrahlung radiation of a fast
         charged particle on a Thomas – Fermi atom . . 103

6.2.   Fast charged particle polarization BR cross
         section on ions in a plasma 108

6.3   Interference – polarization effects during
         radiation of relativistic particles in a dense . . 119

7.   Polarization – interference phenomena in radiation
     of thermal electrons in a low-temperature plasma . . . 121

8.   Polarization radiation and absorption in a laser field . 131

8.1.   Multiphoton polarization bremsstrahlung
         emission and absorption 131

8.2.   Polarization – interference effects in collisions
         of an electron with atoms and ions
         in a near-resonance laser field 135

9.   Polarization radiation, Compton scattering
     and collisional ionization. Cross section relationship,
     similarity laws, new ionization cross section data . . . 150

9.1.   Approximate scaling of the atomic Compton
         profile 150

9.2.   Collisional ionization of atoms. The cross
       section calculation in the Born – Compton
       approximation 153
9.3.   Polarization bremsstrahlung radiation
       with ionization of atom: relation with X-ray
       Compton scattering 164

10. Experimental aspects 173
     10.1.   Experiments on PBR of electrons on atoms . . 173
     10.2.   Relativistic experiments on accelerators 176
     10.3.   Polarization bremsstrahlung of heavy charged
             particles 179
     10.4.   Experiments on the laser-assisted electron
             scattering on atoms 182
     10.5.   Collision-induced absorption in gases 184
     10.6.   Polarization effects near the $4f$-structure
             in BR on metallic targets 190
11. Conclusion 194
References   197

ASYMPTOTIC THEORY OF CHARGE EXCHANGE AND MOBILITY PROCESSES FOR ATOMIC IONS

B. M. Smirnov

1.   Introduction 207
2.   Asymptotic theory of the interaction of atomic ions
     with parent atoms at large separations 209
     2.1.   Character of the resonant charge exchange
            process 209
     2.2.   Ion – atom exchange interaction potential . . . 212
     2.3   Ion – atom exchange interaction for light atoms 218
     2.4.   Ion – atom exchange interaction for heavy
            atoms 222

3.   Asymptotic theory of resonant charge exchange
      process 226
      3.1.   Cross section of resonance charge exchange
              with transition of s-electron 226
      3.2.   Cross section of resonant charge exchange
              with p-electron transition 229
      3.3.   Resonant harge exchange for different cases
              of Hund coupling 234
      3.4.   Average cross sections of resonant charge
              exchange 241
      3.5.   Resonant charge exchange at ultralow energies 243
4.   Mobility of atomic ions in gases 248
      4.1.   The character of ion drift in atomic gas 248
      4.2.   Mobility of ions at low field strengths 253
      4.3.   Mobility of ions in parent gases at low fields . . 256
      4.4.   Mobility of ions at intermediate and high field
              strengths 265
      4.5.   Diffusion of atomic ions in gases in external
              fields 272
5.   Conclusions 278
References   279

PLASMA MODELS OF ATOM AND RADIATIVE – COLLISIONAL PROCESSES

V. A. Astapenko, L. A. Bureyeva, V. S. Lisitsa

1. Introduction

Plasma models of atoms (PMA) have been attractive in atomic physics for many years in spite of the rapid development of computational methods. An obvious advantage of these models is in their simplicity and versatility which allow the description of many properties of complex atoms and ions on a common base. These include the interaction potentials of atoms with charged particles [1, 2], photoionization cross sections of atoms [3, 4], static and dynamic polarizabilities [1, 5 – 8] and other properties.

In recent years, the interest in plasma models increased in connection with intensive research of new channels of bremsstrahlung radiation in collision of atoms with charged (and neutral, in some cases) particles. These channels are caused by the dynamic polarization of atomic (or ionic) cores producing radiation during collisions, the so-called polarization radiation (PR), see [9 – 11]. It is in such processes that plasma properties of atoms appear. The present review is devoted to analysis of atomic plasma models from the viewpoint of their application to the PR processes.

We should note that PMA are basically classical, because due to the Pauli principle most plasma electrons in a complex atom occupy states with large orbital momentum values. In this sense the properties of atom – electron clots can be considered on the base of the same approaches as are applied to fluctuations in plasmas, in particular, to the calculation of dynamic properties of the electron "coat" that screens the Coulomb field around an ion in plasma [12, 13].

1

From this point of view, the atomic nucleus in plasma exhibits a sort of double screening produced by bound electrons of the ion's core and by free plasma electrons that cause the Debye screening. Using PMA, both types of screening can be considered on a common base. Then the problem of collective properties of atomic electron plasma which was discussed in [14, 15] is solved, in our opinion, by an effective description of the specific phenomena utilizing plasma models (see also [6, 7]).

The starting point for describing atomic features in the plasma approximation is the static Thomas – Fermi model and its modifications [1, 2]. The model itself is essentially the simplest plasma model describing atomic properties. Indeed, the Thomas – Fermi distribution for a multielectron atom can be obtained, according to [7], by solving the self-consistent Vlasov's equations traditionally exploited in plasma physics [12 – 16]. The corresponding system of equations has the form (here we use atomic units with $\hbar = e = m_e = 1$):

$$\frac{\partial f}{\partial t} + \mathbf{p}\,\frac{\partial f}{\partial \mathbf{r}} - \nabla U\,\frac{\partial f}{\partial \mathbf{p}} = 0\,, \tag{1.1}$$

$$\Delta U = 4\pi[Z\,\delta(\mathbf{r}) - n(\mathbf{r})]\,, \tag{1.2}$$

$$n(\mathbf{r}, t) = \int f(\mathbf{r}, \mathbf{p}, t)\,d\mathbf{p}\,. \tag{1.3}$$

Here $f(\mathbf{r}, \mathbf{p}, t)$ is the distribution function of electrons, $U(\mathbf{r}, t)$ is the potential energy of electrons in the self-consistent field, $n(\mathbf{r}, t)$ is the electron density distribution, and Z is the charge of the atomic nucleus. In the absence of external electromagnetic fields the distribution of electrons and their energy are $f(\mathbf{r}, \mathbf{p}, t) = f_0(r, p)$, $U(\mathbf{r}, t) = \varphi(r)$, $n(\mathbf{r}, t) = n_0(r)$ respectively. Thus the solution to Eqn. (1.1) can be represented in the form:

$$f_0(r, p) = \frac{2}{(2\pi)^3}\,\theta(E_{\mathrm{F}} - E)\,, \quad E = p^2\big/2 + \varphi(r)\,. \tag{1.4}$$

Here the Fermi energy for degenerate gas of atomic electrons which obey the Pauli principle has been taken into account. Substituting (1.4) into Eqns (1.2) and (1.3) yields the Thomas – Fermi distribution:

$$n_0(r) = \frac{p_{\mathrm{F}}^3}{3\,\pi^2}\,, \quad p_{\mathrm{F}}(r) = \sqrt{2\,(E_{\mathrm{F}} - \varphi(r))}\,, \tag{1.5}$$

$$E_{\mathrm{F}} - \varphi(r) = \frac{Z}{r} \, \chi \left(\frac{r}{r_{\mathrm{TF}}} \right) \, ,$$

$$r_{\mathrm{TF}} = \frac{b}{\sqrt[3]{Z}} \, , \qquad b = \sqrt[3]{\frac{9 \, \pi^2}{128}} \cong 0.8853 \, .$$

(1.6)

Here $\chi(x)$ is the Thomas–Fermi function and r_{TF} is the Thomas–Fermi radius.

Thus, using the atomic plasma model with the additional assumption of the relevance of the Fermi boundary energy to atomic electrons allows us to recover the Thomas–Fermi distribution.

Atomic plasma models permit us to examine, as noted above, the general properties of polarization radiation caused by dynamic polarization of shells of atoms or complex ions. The term "polarization" is used below in the sense of the dynamic polarization of a medium by alternating electric fields of charged particles, typical for plasma applications. However, the role of the medium in atomic processes considered below is provided by a complex atom with its own electrons. It is these electrons that are polarized by the incident charged (or neutral) particle.

The polarization radiation can be most easily understood by its analogy with the light scattering by atoms. The physical analogy of the polarization effects can be clearly illustrated. Figure 1 shows the classical scheme of the polarization radiation (Fig. 1a) and Feynmann's diagrams (Fig. 1b) which describe within the frames of the first Born approximation PR of electrons in collision with atoms ("inelastic" or non-coherent PR). If the target's core does not change its state ("elastic" or coherent PR) then, naturally, $f = i$. The same diagrams describe polarization effects (1) in photoionization if the initial state of the incident electron is substituted by the bound state, (2) in photorecombination if the final state is considered as bound and the initial state as free, and, finally, (3) in bound–bound transitions if both states are considered bound. The double line connecting the vertexes of the displayed diagrams represents the electron propagator of the target core, which in the case of the "elastic" PR is expressed in terms of the dynamic polarizability and for the bremsstrahlung — in terms of the Compton scattering cross section.

It is important to notice that in the case of the polarization radiation the perturbed atom is the emitting object while the energy is being lost by the perturbing particle (electron or ion). The process looks inelastic with respect to the perturbing particle although the photon is emitted by the atomic frame. It is this last fact that causes the polarization radiation to be independent of the incident particle mass, in contrast with the static channel.

For free – free transitions, the process of PR is similar to the ordinary bremsstrahlung, hence the term polarization bremsstrahlung, PBR [9]. For bound – bound transitions, PR processes are mostly known as the core polarization effects changing oscillator strengths of radiative transitions. Clearly, free-bound transitions can be called polarization recombination [17].

The subdivision of the emitting systems on the perturbing and polarized parts is rather conditional and is mainly used to make representation clear. In fact, the entire compound system "atom + perturbing particles" produces radiation. Such an approach to the problem of radiation by the compound was first suggested by M. Born in 1940 when he studied the general theory of bremsstrahlung (see Chapt. 22, p. 9 in the book by Mott and Messi [18]). This approach entails, in particular, the possibility of interference of the ordinary and polarization radiative channels. The ordinary channel corresponding to the motion of the perturbing particle in the potential of the "frozen-out" non-perturbed atom will be called static below, in contrast to the above-mentioned polarization channel corresponding to deformation of an atom (or its core).

One of the first estimates of the polarization channel in radiation were done by S. P. Kapitza [19] and M. L. Ter-Mikaelyan [20] within the frames of purely classical electrodynamics. These calculations were based on the notion that scattering can occur on the proper electric field of a charged particle inside matter, which can be taken into account by multiplying the intensity of this field by the extinction coefficient in this medium. By expressing this coefficient in terms of the static polarizability of the medium one can obtain the simple relation (see [21], the problem to §119):

$$dI(\omega) = \frac{8\,\pi\,\omega^4\,\sqrt{\varepsilon}\,N\,\alpha^2}{3\,c^3}\int |E(\omega)|^2\,dV\,d\omega \qquad (1.7)$$

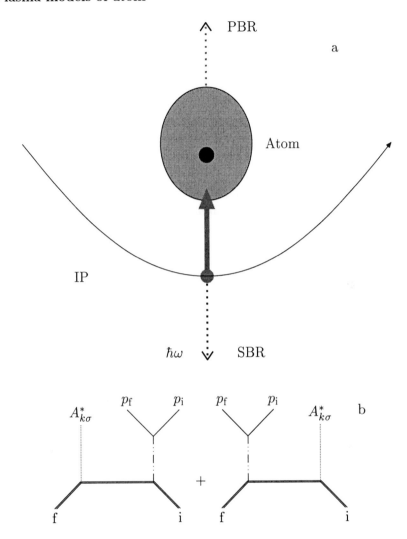

Fig. 1. (a) The classical scheme of the polarization BR of a charged particle on an atom. (b) Feynmann's diagrams describing the polarization BR of an electron on an atom in the first Born approximation. The double lines relate to the electron core of the atom, the single lines mark the incident electron, the dashed line shows a free electromagnetic field, and the dashed-dotted line denotes the photon propagator.

where ε is the dielectric permeability, N, α are the number density and the polarizability of the scattering centers of the medium, respectively, $E(\omega)$ is the Fourier-image of the charged particle's field at frequency ω, c is the speed of light.

The approach described above stems from the well-known method of equivalent photons proposed by Fermi [22], who treated interaction of charged particles with atoms as absorptions of an equivalent photon flux with intensity $I(\omega)$ determined by the Fourier-decomposition of the electric field generated by the charged particle. Note that the photon equivalence is the exact result for relativistic particles (the Weizsecker – Williams theorem), but below we use an approximate version of this theorem related to the field of non-relativistic particles, as it was the case in the original Fermi's work.

This field can be found from equations of motion of the incident electron in the atomic potential

$$\mathbf{F} = e\mathbf{r}/r^3 = -m_e\ddot{\mathbf{r}}/Z_{\mathrm{ef}}e, \quad \mathbf{F}_\omega = -m_e\omega^2\mathbf{d}_\omega/Z_{\mathrm{ef}}e^2. \tag{1.8}$$

Here Z_{ef} is the effective charge of the ionic core (depending upon the electron energy and the emitting frequency).

Therefore, it is easy to generalize the above results on the case of the dynamic polarization of an atom by a particle's field by multiplying the intensity $I(\omega)$, determined by the square of the modulus of Fourier-component (1.8), with the scattering cross section $\sigma_{\mathrm{scat}}(\omega)$:

$$I^{\mathrm{PR}}(\omega) = \sigma_{\mathrm{scat}}(\omega) I(\omega). \tag{1.9}$$

The cross section $\sigma_{\mathrm{scat}}(\omega)$ is connected to the dynamic polarizability of an ion (atom) $\alpha(\omega)$ at the frequency ω:

$$\sigma_{\mathrm{scat}}(\omega) = \frac{8\pi\omega^4}{3c^4}|\alpha(\omega)|^2, \tag{1.10}$$

$$\alpha(\omega) = \sum_n |d_{in}|^2 \left[(\omega_{ni} - \omega - i0)^{-1} + (\omega_{ni} + \omega - i0)^{-1}\right], \tag{1.11}$$

in which (due to non-resonance conditions) one should take into account (virtual) transitions of the emitting electron into all other energy levels of the ion (enumerated in Eqn. (1.11) by index n).

Making use of relationships (1.8) allows us to easily connect the PR intensity with the bremsstrahlung intensity at some effective Coulomb center with charge Z_{ef}. Indeed, the particle's equation of motion ensues that the field induced by the particle itself at the nucleus is Z_{ef} times as small as the nuclear field that is responsible for acceleration of the electron and thus its radiation. From here the simple expression for the ratio between the incident particle scattered field intensity and the radiated field intensity during interaction with the nucleus can be found:

$$\frac{I^{PR}(\omega)}{I^{Br}\omega} = \left[\frac{m\omega^2|\alpha(\omega)|}{e^2 Z_{ef}}\right]^2 \equiv R(\omega) \qquad (1.12)$$

This expression demonstrates that the spectral intensity distribution for the incident particle falls off the relationship.

Expression (1.12) corresponds to the classical dipole approximation for the charged particle interaction with atoms of matter and, in spite of its apparent simplicity, describes a number of important relationships for PR. This expression implies that the polarization effects are proportional, naturally, to the polarizability of matter. At small frequencies, the dynamical polarizability transforms into the static one and polarization effects rapidly decrease with decreasing frequency. At large frequencies, the dynamical polarizability corresponds to scattering on quasi-free electrons $\alpha(\omega) = -N_{ef} e^2/m\omega^2$, so that the ratio of both channels is equal to the ratio of the number of bound electrons N_{ef} to the effective nuclear charge of the atomic core Z_{ef}:

$$R(\omega) = \left(\frac{N_{ef}}{Z_{ef}}\right)^2. \qquad (1.13)$$

It is clear that the effective charge Z_{ef}, as well as the effective number of quasi-free electrons N_{ef}, depends upon the penetration degree of the incident electron into the atomic core, the correct allowance for which is essential to calculate the contribution from the polarization channel into radiation emission.

The above concept of PR as scattering of the incident particle's proper field suggests a broad analogy between the PR and Compton scattering by atoms [23], and also between the latter process and the collisional ionization of atoms by electrons, see [18]. Such analogs allow many results relating to one field to be applied to another one and thus to obtain some new data.

The structure of the review is as follows. In Chapter 2 we present the general results of atomic plasma theory and compare it with more rigorous quantum calculations. Chapter 3 is devoted to the general theory of radiation emission of charged particles scattering on complex ions in plasma, accounting for polarization interaction channels both with the ionic core and the Debye "coat". In Chapter 4 we review atomic plasma model applications for photoionization calculations. Chapter 5 analyzes bremsstrahlung and photorecombination radiation of charged particles in plasma with complex ions and estimates the polarization channel contributions into the total recombination rate and plasma radiative emission efficiency. The polarization channels of radiation emission of fast particles in plasma are considered in Chap. 6. The polarization channel contribution to a low-temperature plasma radiation emission with account for its interference with the static channel are presented in Chap. 7. Multiphoton emission – absorption polarization processes in a laser radiation are considered in Chap. 8. In Chapter 9, we use the interrelation between PR processes, the Compton scattering, and collisional ionization for a more accurate calculation of the ionization cross sections of complex atoms. Chapter 10 concerns with experiments on observing polarization radiation emission in various media.

2. Plasma characteristics of atoms in phenomenological and kinetic models of atoms. Dynamic polarizability of atomic structures

A key feature describing polarization effects in atomic transitions is dynamic (in the general case), generalized (non-dipole) polarizability of an atom (ion) or atom's core $\alpha(\omega, q)$. Its calculation is a complicated task which allows an exact solution only for hydrogen-like atoms [24]. Quantum mechanical calculations of the dynamic polarizability of multielectron atoms are sufficiently difficult and laborious. At the same time, there are model approaches to handle this problem which are based rather on the physical intuition than on utilizing a consistent mathematical formalism. The local plasma frequency (LPF) approximation, or the Brandt–Lundqvist model [3], is one such simple model.

2.1. Brandt–Lundqvist local plasma frequency model

The local plasma model was suggested in paper [3] to describe the photoabsorption by many-electron atoms in the spectral range $\omega \sim Z$ Ry. In contrast to the high ($\omega \sim Z^2$ Ry) and low ($\omega \sim$ Ry) frequency ranges, absorption of a photon at these intermediate frequencies is mainly determined by collective effects and to the less extent by single-particle interactions. Based on these qualitative considerations, the electronic core of an atom is approximated by an inhomogeneous charge distribution whose interaction with electromagnetic field occurs via the plasma resonance (atomic units — au):

$$\omega = \omega_{\mathrm{p}}(r) = \sqrt{4\pi\,n(r)}\,. \tag{2.1}$$

Here $n(r)$ is the local electron density and $\omega_{\mathrm{p}}(r)$ is the corresponding local plasma frequency. It is easy to check that condition (2.1) entails the following expression for the dipole dynamic polarizability that respects the dispersion relations and the sum rules [3]:

$$\alpha^{\mathrm{BL}}(\omega) = \int\limits_0^{R_0} \frac{\omega_{\mathrm{p}}^2(r)\,r^2\,dr}{\omega_{\mathrm{p}}^2(r) - \omega^2 - i0} = \int \alpha^{\mathrm{BL}}(r,\omega)\,d\mathbf{r}\,. \tag{2.2}$$

Here we introduced the quantity $\alpha^{\mathrm{BL}}(r,\omega)$ that can be naturally named the polarizability spatial density in the Brandt–Lundqvist approximation, R_0 is the size of the atom (ion).

The generalized polarizability in the LPF model can also be conveniently written in the coordinate representation $\alpha(\omega, r)$. The corresponding expression [25] differs from formula (2.2) only by the upper limit of integration:

$$\alpha(\omega, r) = \int_0^r \frac{\omega_p^2(r')\, r'^2\, dr'}{\omega_p^2(r') - \omega^2 - i0}, \quad \alpha(\omega, q) = \int_0^\infty \alpha(\omega, r)\, j_0(qr)\, r^2 dr.$$

(2.2a)

where $j_0(x)$ is the zero-order spherical Bessel function.

Expression (2.2) has the correct high-frequency asymptotic

$$\alpha^{BL}(\omega \to \infty) \to -\frac{N_e}{\omega^2},$$

(2.3)

N_e is the number of electrons in the atom. In the low frequency limit Eqn. (2.2a) yields:

$$\alpha^{BL}(\omega \to 0) \to R_0^3 / 3.$$

(2.4)

Despite the apparent simplicity, formula (2.4) in some cases describes well the existing experimental data. This firstly relates to multielectron atoms with closed shells because then the main contribution to the polarizability is provided by the atom's bound energy spectrum [26] and the local plasma frequency approximation is most adequate. This fact is illustrated by the following table:

Table 1.

Atom (ion)	ArI	KrI	XeI	KII	RbII	CsII	SrIII	BaIII
α_0^{exp} (a.u.)	11	17	27	7.5	12	16.3	6.6	11.4
α_0^{var} (a.u.)	19.3	26.8	30.9	9.1	14.3	17.8	8.7	11.4
α_0^{VSh} (a.u.)		21.1	25.5	6.6	11.9	15.3	7.5	9.7
α_0^{SZ} (a.u.)	11.6	17.2	27.3	5.25	8.5	14.6		
α_0^{BL} (a.u.)	22	24	27	8.6	11.6	13.5	7	8.4

Here α_0^{var} means computations by the variation method [1], α_0^{VSh} stands for Vinogradov and Shevelko's calculations [5] (see Sect. 2.2), α_0^{SZ} is Stott and Zaremba's calculations [27] utilizing the electron density formalism, α_0^{BL} denotes computations in the Brandt – Lundqvist model. Calculations of the static polarizability in the Brandt – Lundqvist model made use of the radius of an atom (ion) computed with account of the correlation correction in the Thomas – Fermi – Dirac model.

Table 1 implies that the static polarizability of atoms (ions) with completed shells calculated by the BL method is in fair agreement with experiments.

Note that expression (2.2) represents the simplest realization of the electron density functional (EDF) formalism for polarizability. Much more complicated realizations of EDF within the quantum mechanical approach frames are reported in papers [27] and [28]. Note that in [27] the static polarizability of spherically symmetric systems is expressed through the polarizability spatial density:

$$\alpha_0 = \frac{4}{3}\pi \int\limits_0^\infty r^3 \alpha(r)\,dr\,. \tag{2.5}$$

In this expression, however, function $\alpha(r)$ is a solution of a complex integral equation.

Different modifications of basic formula (2.2) were considered in relation with the problem of computation of atom photoeffect cross sections [29]. These modifications were based on considering, in addition to the local plasma frequency (2.1), the local "single-particle" frequency:

$$\omega_{\text{sp}}^2(r) = \frac{1}{r^3}N_{\text{out}}(r)\,, \quad N_{\text{out}}(r) = \int\limits_r^\infty n(r')\,d^3r' \tag{2.6}$$

and an atom's dielectric permeability in the simplest form:

$$\varepsilon(r,\omega) = 1 - \frac{\omega_{\text{p}}^2(r)}{\omega^2}\,. \tag{2.7}$$

One expression used in [29] for the dynamic polarizability is the equality

$$\alpha(\omega) = \int d^3r\,\frac{n(r)\,\varepsilon^{-1}(r,\omega)}{\omega_{\text{sp}}^2 - \omega^2}\,. \tag{2.8}$$

Using Eqns (2.2) and (2.8) yields close results. The model used for the local atom's electron density has more effect on the results. For a Thomas–Fermi atom it is possible to obtain the expression for the dynamic polarizability that reveals the similarity (scaling) law with respect to the ratio of the frequency to the nuclear charge:

$$\alpha(\omega,\, Z) = r_{\mathrm{TF}}^3\, \beta\left(\frac{\omega}{Z}\right) = \frac{b^3\, a_0^3}{Z}\, \beta\left(\frac{\omega}{Z}\right),$$

$$\beta(\nu) = \int\limits_0^{x_0} \frac{4\,\pi\, f(x)\, x^2\, dx}{4\,\pi\, f(x) - \nu^2 - i0}.$$

(2.9)

Here $r_{\mathrm{TF}} = b\, a_0/Z^{1/3}$ is the Thomas–Fermi radius,

$$b = \left(9\pi^2/128\right)^{1/3} \cong 0.8853,$$

Z is the atomic nuclear charge, a_0 the Bohr radius, $\beta(\nu)$ is the dimensionless polarizability as a function of the reduced frequency $\nu = \omega/Z$, $x_0 = R_0/r_{\mathrm{TF}}$ is the reduced atomic radius,

$$f(x) = f_{\mathrm{TF}}(x) = \frac{32}{9\,\pi^3} \left(\frac{\chi(x)}{x}\right)^{3/2}$$

(2.10)

$\chi(x)$ is the Thomas–Fermi function.

Note that formulas (2.9) are also applicable for other statistic models of the electron density, in which it can be presented in the form:

$$n(r) = Z^2\, f\left(x = r/r_{\mathrm{TF}}\right).$$

(2.11)

The similarity function $f(x)$ in Eqn. (2.9), instead of (2.10), can be more conveniently taken from the Lenz–Jensen model [1]:

$$f_{\mathrm{LJ}}(x) \cong 3.7\, e^{-\sqrt{9.7\,x}}\, \frac{\left(1 + 0.26\,\sqrt{9.7\,x}\right)^3}{(9.7\,x)^{3/2}}.$$

(2.12)

Lenz–Jensen function (2.12) is close to its analog (2.10) in the Thomas–Fermi model; an advantage is that it has a more realistic behavior at large x.

The results of calculations of the real and imaginary parts of the dipole polarizability of the krypton atom using the local plasma frequency model (with the Slater and Lenz – Jensen electron density) are presented in Fig. 2. In the same figure we show the results of calculations of the corresponding values in the random phases with exchange approximation obtained in paper [30]. It is clear that the dynamic polarizability of the krypton atom computed in the LPF model with the Lenz – Jensen electron density smoothly reproduces quantum-mechanical features of the frequency behavior. These appear most obviously near the ionization potentials of electronic subshells, in a way like the statistical density of the electron distribution reproduces the exact quantum-mechanical relationship. The usage of the Slater wave functions in this model reveals, to some degree, spectral oscillations of the polarizability near the ionization potentials of electronic subshells. Here, however, the universality of description, featured to the statistical model of atom, breaks down. Despite a qualitative character, an advantage of the Brandt – Lundqvist approximation is its simplicity, clearness, and universality.

Let us consider a model example in which the usage of the Brandt – Lundqvist approximation allows an explicit expression for the complex polarizability to be obtained. It corresponds to an inhomogeneous plasma formation with linear electron density distribution along radius, when the function $f(x)$ in Eqn. (2.11) can be represented in the form

$$f_{\text{lin}}(x) = \frac{128}{3\,\pi^3\,x_0^3}\left(1 - \frac{x}{x_0}\right), \quad x \le x_0. \tag{2.13}$$

Then Eqn. (2.19) yields:

$$\text{Im}\{\beta(\nu)\} = \frac{3\,\pi^3}{512}\,x_0^6\,\nu^2\left(1 - \frac{3\,\pi^2}{512}\,x_0^3\,\nu^2\right), \quad \nu \le \sqrt{\frac{512}{3\,\pi^2\,x_0^3}}, \tag{2.14}$$

$$\text{Re}\{\beta(\nu)\} = x_0^3\,I\left(\sqrt{\frac{3\,\pi^2\,x_0^3}{512}}\,\nu\right),$$

$$I(y) = \frac{1}{3} - \frac{3}{2}\,y^2 + y^4 + y^2\,(y^2 - 1)\,\ln\frac{|1 - y^2|}{y^2}. \tag{2.15}$$

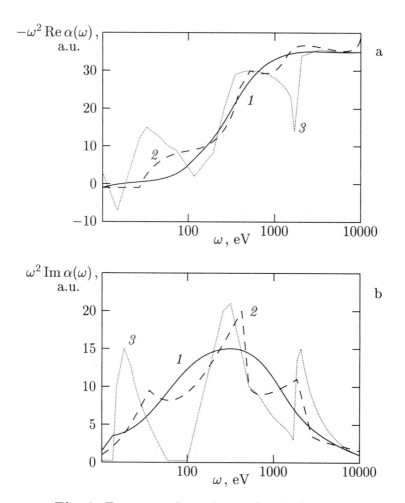

Fig. 2. Frequency dependences (multiplied by the frequency square) of the real (a) and imaginary (b) parts of the krypton atom polarizability calculated in different approximations: *1* — the local plasma density for the Lenz – Jensen electron density, *2* — the local plasma density for the Slater electron density, *3* — random phase with exchange approximation [30].

Parameter x_0, which is the reduced radius of the plasma formation, can be evaluated if the static polarizability α_0 is known using the formula:

$$x_0 = 4 \left(\frac{2\,Z\,\alpha_0}{3\,\pi^2} \right)^{1/3} . \tag{2.16}$$

Note that the reduced radius of ion has been calculated in paper [5] within the Thomas – Fermi – Dirac model frames as a function of the ionization degree $q = (Z - N_e)/Z$, where the following expression has been obtained for $x_0(q)$:

$$x_0(q) \approx \begin{cases} 6.84\,q^{-1/3} , & q \leq 0.05 , \\ 2.96 \left(\frac{1-q}{q} \right)^{2/3} , & 0.2 < q \leq 1 . \end{cases} \tag{2.17}$$

A comparison of the frequency dependence of the real part of polarizability (2.15) with the parameter $x_0 = 9.2$ (which would correspond to the argon atom if one uses Eqn. (2.16) to find the reduced radius), with results of a quantum-mechanical computation in the approximation of random phases with exchange (RPEA) carried out in [26] for the argon atom, indicates that the linear approximation for the atomic electron density is too crude. The most appreciable difference of these calculations is that the RPEA yields the real part of the dynamic polarizability, compressed approximately two times along the abscissa compared to predictions of the simplified approach described above.

2.2. The response functions and the static polarizability of a Thomas – Fermi atom

There are more rigorous approaches to compute the polarizability of a Thomas – Fermi atom. For example, in paper [5] the calculation of the static polarizability α_0 of the atom (ion) is based on finding atom's response on an external electric field. The problem is reduced to solving a differential equiation for the induced potential $\varphi_1(r)$ using the Thomas – Fermi approximation. The corresponding equation for the electric potential induced by an external field with strength $\mathbf{E}$ has the form:

$$\Delta\varphi_1(\mathbf{r}) = \frac{4\sqrt{2}}{\pi} \left(\varphi_0(r) - \frac{Z - N_e}{r_0} \right)^{1/2} \varphi_1(\mathbf{r}), \quad r < r_0 . \tag{2.18}$$

Here φ_0 is the unperturbed potential and r_0 is the atomic size. Equality (2.18) results from linearization of the initial Thomas–Fermi equation for the total potential.

Outside the atom, the induced potential is expressed in terms of atom's static polarizability determined by the formula:

$$\varphi_1(\mathbf{r}) = -\mathbf{r}\mathbf{E} + \alpha_0 \frac{\mathbf{E}\mathbf{r}}{r^3}, \quad r > r_0. \tag{2.19}$$

To separate angular variable, a new function $u(r)$ is introduced:

$$\varphi_1(\mathbf{r}) = (\mathbf{E}\mathbf{r})\, u(r). \tag{2.20}$$

Making use of the continuity condition for $u(r)$ and its derivative u' at the atomic boundary, the following expression was obtained in [5] for the static polarizability:

$$\alpha_0 = r_0^3 \left(1 + \frac{3\, u(r_0)}{r_0\, u'(r_0)}\right)^{-1}. \tag{2.21}$$

In the case of multicharged ions $Z \gg N_e$, when one sets $u(r) \approx -1$ in the first approximation, the static polarizability is found to be:

$$\alpha_0 = \frac{63}{16} \frac{N_e^3}{Z^4}. \tag{2.22}$$

The dependence of the static polarizability of a multicharged ion on the nuclear charge, which is proportional to Z^{-4} and following from Eqn. (2.22), can also be derived from a more rigorous quantum treatment for ions with completed shells. Indeed, in that case, if $Z \gg N_e$, the minimum frequency of virtual transition is proportional to the nuclear charge square. Then the above dependence on Z follows straightforwardly from the general quantum-mechanical expression for the polarizability:

$$\alpha(\omega) = \sum_n \frac{f_{in}}{\omega_{in}^2 - \omega^2}. \tag{2.23}$$

Note that for a hydrogen-like ion the static polarizability takes the exact value:

$$\alpha_0 = \frac{9}{2} \frac{1}{Z^4}, \tag{2.24}$$

whereas the usage of Eqn. (2.22) gives rise to the 14% error. For non-completed electronic shells the main contribution to the static polarizability is provided by the virtual transition without change of the main quantum number. In this case $\alpha_0 \propto Z^{-3}$.

A similar method can be applied to calculate static multipole polarizability of atoms and ions in the Thomas–Fermi model [8]. The general expression for the multipole polarizability α_κ has the same structure as Eqn. (2.21). A simple analytical representation for α_κ can be found in the case of multicharge ions [8]:

$$\alpha_\kappa = \left(\frac{3^{4\kappa+5}}{2^{4\kappa-1}}\right)^{1/3} \frac{(4\kappa+3)!! \, N^{(4\kappa+5)/3} \, Z^{-(2\kappa+2)}}{(2\kappa+1)\, \Gamma(2\kappa+4)} a_0^{2\kappa+1}, \quad Z \gg N_e. \tag{2.25}$$

To conclude this point, we note that for ions with completed outer shells there is a simple empirical formula for the static dipole polarizaility of the external shell with the main quantum number n that gives fair agreement with experiments:

$$\alpha_0 = N_n \frac{n^6}{Z_n^4}. \tag{2.26}$$

Here N_n and Z_n are the number of electrons in the external shell and the effective nuclear charge, respectively. The latter can be determined as $Z_n = n\sqrt{2I_n}$, where I_n is the external shell ionization potential.

Formula (2.26) gives an especially good result for neon-like atoms ($n = 2$, $N_n = 8$), as follows from Table 2. In this Table we also display the results of caluclations using Eqns (2.21) from [5, 8] and experimental data from [31].

Here the number in round brackets means the power of 10 which should be multiplied by the given value. As is seen from this Table, the agreement with experiment gets somewhat worse with increasing the ionization degree of ions, when the situation becomes more hydrogen-like, and Eqn. (2.26) gives a slightly smaller value. At the same time, using Eqn. (2.26) gives better agreement with experiment than Eqn. (2.21), especially for high ionization degrees.

Table 2.

ion	Ar^{+8}	Ca^{+10}	Ti^{+12}	Fe^{+16}	Co^{+17}
I_n, eV	396.4	558.2	737.8	1168	1293
Z_n	10.8	12.8	14.73	18.54	19.5
α_0^{exper}	3.16(-2)	1.74(-2)	1.04(-2)	4.44(-3)	3.69(-3)
α_0 (2.21)	3.75(-2)	2.46(-2)	1.68(-2)	8.61(-3)	7.41(-3)
α_0 (2.26)	3.76(-2)	1.89(-2)	1.09(-2)	4.33(-3)	3.53(-3)

ion	Ni^{+18}	Zn^{+20}	Kr^{+26}	Mo^{+32}
I_n, eV	1419	1693	2728	3960
Z_n	20.43	22.32	28.33	34.13
α_0^{exper}	3.08(-3)	2.63(-3)	9.31(-4)	4.62(-4)
α_0 (2.21)	6.4(-3)	10(-3)	4.46(-3)	2.24(-3)
α_0 (2.26)	2.94(-3)	2.06(-3)	7.95(-4)	3.77(-4)

2.3. Kinetic model of the dynamic polarizability of atoms

A more general approach was developed in [7] to calculate the dynamic polarizability of atoms. In that paper, initial equations were taken in the form of Vlasov's kinetic equations for the self-consistent field. The boundary conditions were chosen to correspond to the problem of radiation scattering on the atomic clot. These boundary conditions differ from those with constant parameter values on the boundary [14] corresponding to the problem of the proper oscillations of atom's electronic core. So no sharp resonances at the plasma frequency appear in the method [7] and the general setting of the problem is more adequate to the problem of the dynamic polarizability calculation. The solution of this scattering problem reduces to calculating the electric potential $\varphi(r)$ in an inhomogeneous medium which is described by the equation

$$\text{div}[\varepsilon(\mathbf{r}, \omega)\, \nabla\varphi(\mathbf{r})] = 0 \qquad (2.27)$$

with boundary conditions at infinity (2.19) and the local dielectric permeability (2.7). Note that Eqn. (2.27) is valid in the long-wavelength approximation $\lambda \gg R_0$, which, nonetheless, does not contradict to high-frequency approximation for the dielectric permeability (2.7). Indeed, the conditions for the both approximations are satisfied simultaneously (for moderate nuclear charges and polarization degrees) when the frequency lies in the broad range $10\,\text{eV} \ll \omega \ll 10\,\text{keV}$. As in the Brandt–Lundqvist approximation, the usage of the Thomas–Fermi model for the electron density of atom (ion) leads to a scaling law for the polarizability like in Eqn. (2.9):

$$\alpha(\omega,\, Z,\, q) = \frac{1}{Z}\, \alpha_X\left(\frac{\omega}{Z},\, q\right). \qquad (2.28)$$

Here $q = (Z - N_e)/Z$ is the ionization multiplicity, α_X is a universal function of the reduced frequency $\nu = \omega/Z$ calculated in [7]. Note that within the frames of the approach discussed the static polarizability is found to be

$$\text{Re}\{\alpha_X(\nu = 0)\} = x_0^3 \qquad (2.29)$$

as opposite to the Brandt–Lundqvist model predictions (2.4). Result (2.29) noticeably overestimates the static polarizability, although it is derived from a more rigorous theory. For example, the maximum of the imaginary part of the polarizability α_X occurs (for $q = 0.3$) at the reduced frequency $\nu_{\max} = 0.06$ a.u. This is by almost an order of magnitude lower than in the local plasma frequency approximation for a neutral Thomas–Fermi atom.

The Brandt–Lundqvist approximation follows, as was noted in [7], from the corresponding solution of Eqn. (2.27) if the term $\propto \varepsilon'/\varepsilon$ is taken into account using perturbation theory, which, generally speaking, violates self-consistency of the problem. Nevertheless, the use of the more simple expression (2.2) in numerical calculations of the dynamic polarizability of atom seems to be preferable.

Note that in some cases Eqn. (2.27) with boundary condition (2.19) allows an analytical solution and the expression for the dynamic polarizability can be written explicitly. In paper [6] this problem was solved for a model case of an inhomogeneous dielectric particle with electron density changing along the radius:

$$n(r) = n_0 \left(1 - r/a\right) . \tag{2.30}$$

The solution implies that at frequencies $\omega > \omega_0 = \sqrt{4\pi\, n_0}$ the polarizability is real and at $\omega < \omega_0$ $\alpha(\omega)$ becomes a complex value, with the imaginary part reaching a maximum at the frequency $\omega = 0.273\,\omega_0$ with the width $\gamma = 0.076\,\omega_0$. The function $\alpha(\omega)$ having imaginary part is an interesting distinctive feature of the inhomogeneous dielectric particle, because a homogeneous particle with real dielectric permeability can not absorb radiation at all.

2.4. Quantum calculations of the dynamic polarizability

For the hydrogen atom and hydrogen-like ions the problem of determination of the dynamic polarizability can be exactly solved (see, for example, [32]). For an atom in the $1s$-state, the dynamic polarizability can be obtained using the Coulomb Green function formalism to have the form:

$$\alpha_{1s}(\omega) = -\frac{1}{\omega^2}\left\{1 - T(E_{1s} + \omega) - T(E_{1s} - \omega)\right\},$$

$$T(E) = \frac{2^7\,\eta^5}{(2 - \eta)\,(1 + \eta)^8}\, F\!\left[2 - \eta, 4, 3; \left(\frac{1 - \eta}{1 + \eta}\right)^2\right], \quad \eta = \frac{Z}{\sqrt{-2\,E}} .$$
$$\tag{2.31}$$

Here Z is the nuclear charge, $F(z)$ is the hypergeometric function.

The generalized (non-dipole) polarizability of a hydrogen-like atom can be represented in a closed form as well. The corresponding expression for the ground state follows from matrix elements obtained in [24] using Green function methods and in [33] employing the variational principle. For excited states the expression in [30] was obtained with the help of the Green function method in the coordinate representation.

For atomic systems in which many-particle effects are weakly expressed in the polarizability at frequencies considered, for example at frequencies of the order of the first ionization potential for the isoelectron row of the alkaline metals, the polarizability can be represented in the hydrogen-like form using the quantum defect method or model potential [32]. In this case, virtual transitions of the external electron in the discrete spectrum which have high oscillation strengths are the main conributors to the polarizability.

In the case of negative ions (for frequencies below or of the order of the photon ionization potential), the main contribution to the dynamic polarizability is provided by the external electron, which weakly interacts not only with electrons of the core, but also with the atomic nucleus. The dynamic polarizability of the negative hydrogen atom is calculated in [34].

In some cases, the single-particle approximation proves to be insufficient to calculate the polarizability due to importance of many-particle effects. This takes place, for example, for atoms with completed shells. Here one should apply methods of the many-particle perturbation theory [15]. The initial expression for the generalized polarizability in such calculations has the form [9, Chap. 7]:

$$\alpha(\omega, q) = \frac{-i}{(\mathbf{eq})} \sum_{\varepsilon > F, j \leq F} \frac{2 \langle j| \exp(i\, \mathbf{qr})|\varepsilon\rangle \langle \varepsilon| \, (\mathbf{eD}(\omega)) \, |j\rangle \, (\varepsilon + I_j)}{(\varepsilon + I_j)^2 - \omega^2},$$

(2.32)

where $\mathbf{e}$ is the unity vector of the radiation polarization, F is the Fermi level, I_j is the ionization potential from j-th shell. The summation is performed over occupied shells and free states including the integration over the continuum. The many-particle correlations in Eqn. (2.32) are taken into account by introducing the effective dipole moment $\mathbf{D}(\omega)$, which is the solution of some integral equation. Using graphical representation, this equation in the approximation of random phases with exchange consists of diagrams that have the creation and annihilation of an electron – hole pair as the principal structure unit. The latter process can be interpreted as the induction of polarization in the target's electronic core. The corrections due to the random phase approximation can appreciably alter the dynamic polarizability and other radiation characteristics

Table 3. Static polarizabilities of the noble gas atoms (in atomic units) calculated by various methods and their experimental values.

Method	Ne	Ar	Kr	Xe
HF(∇) [26]	1.88	7.40	11.15	17.25
HF(r) [26]	2.47	12.39	18.98	32.45
RPEA [26]	2.30	10.73	16.18	27.98
FFP [27]	2.76	10.6	15.7	24.9
Experiment [35]	2.66	11.09	16.75	27.32

of multielectron system, such as photoionization cross section. Here correlations of other types could be significant, as indicated by comparison of the calculated photoeffect cross sections with experimental data [15].

The polarizability of the noble gas atoms in the random phase with exchange approximation were calculated in paper [26]. It was shown that choosing the Hamiltonian in the form of the coordinate or momentum, without taking into account many-particle correlations, leads to significantly different results for the polarization. Making allowance for the correlations, both approaches yield virtually the same values for polarization which are very close to experimental data. It was also established that virtual transitions from the external shell into the d-th states of the continuum mostly contribute to the polarization of noble gas atoms. The contribution from the transition in the discrete spectrum is $10-20\%$. The static polarizability of the noble gas atoms was calculated self-consistently in [27]. The results of these calculations and experimental data for the noble gas atoms are listed in Table 3.

Dynamic polarizability of the Rydberg states with fixed spherical quantum numbers n, l, m was calculated in paper [36] employing quasi-classic expressions for radial integrals which enters the oscillator strengths (see Eqn. (2.23)). These calculations revealed some distinctive features of the polarizability of these states, which are due to approximately equidistant energy spectrum of the Rydberg energy levels. This fact ensues a qualitatively different behavior of the resonance polarizability due to superposition of contributions of the above- and below-lying levels. As a result, the dynamic polarizability of highly excited hydrogen-like states does not change the sign at the resonance crossing, and vanishes either twice or any inside the inter-resonance intervals. The non-resonance dynamic polarizability of such levels is proportional to the sixth power of the principle quantum number, while each separate term in expression (2.23) is proportional to the seventh power which in turn is due to the above mentioned mutual compensation of contributions from the above- and below-lying states. The quasi-classical expression for static polarizability has the form [36]:

$$\alpha_{nlm}(\omega \ll n^{-3}) = 14\, n^6 \left(1 - \frac{m^2}{l^2}\right)$$

$$\times \sum_{s=1}^{\infty} \frac{1}{s^2} \left[J_s^2(s\,\varepsilon) + \frac{1 - \varepsilon^2}{\varepsilon^2} J_s'^2(s\,\varepsilon)\right] , \quad (2.33)$$

where $\varepsilon = \sqrt{1 - (l/\tilde{n})^2}$ is a quantity analogical to the orbital eccentricity of the classical motion of electron, $\tilde{n} = 2\, n\, n'/(n - n')$ is the effective principle quantum number, $J_s(z)$ is the Bessel function. In equation (2.33) the nuclear charge is set to unity. For $m = l = 0$ Eqn. (2.34) yields $\alpha_{n00} \approx 0.6\, n^6$.

For highly excited states of complex atoms, the results obtained remain valid if $l > 2$, when the quantum defect is negligible [36]. In the opposite case of small quantum numbers l, no compensation of contributions to the polarizability for a given highly excited level from above- and below-lying levels occurs. So the dynamic polarizability turns out to be proportional to the seventh power of the principle quantum number. For example, for a spherically symmetric state of an atom ($l = m = 0$) one can obtain the following expression:

$$\alpha_{n00} = \frac{3\, n^7\, \delta}{\delta^2 - (\omega\, n^3)^2} , \qquad \delta = \delta_1 - \delta_0 , \quad (2.34)$$

where $\delta_{0,1}$ are quantum defects of states with $l = 0, 1$.

The dynamic polarizability of the Rydberg states averaged over angular quantum numbers was studied in paper [37] using the

Kramers approximation for oscillator strengths. The general expression for $\langle\alpha(\omega)\rangle_{lm}$ is given via elementary functions. This expression, in particular can be used to derive the simple formula for the static polarizability of a Rydberg atom:

$$\alpha(0) = \frac{15\,n^6 + 21\,n^4}{8\,Z^4}\,. \tag{2.35}$$

To summarize, we can stress the efficiency of the atomic plasma model in calculations of most important characteristics of the atom, such as static and dynamic polarizability which play a key role in radiative-collisional processes with participation of heavy atoms and ions. Of course, as with any statistical model, the plasma model of the atom does not take into account details of the polarization properties which are due to the shell-like atomic structure and other quantum effects. Nevertheless, this model describes satisfactorily the shell-averaged atomic characteristics which quite often are sufficient to consider in the observed plasma effects associated, as a rule, with such characteristics. In view of the approximate character of plasma models, a more rigorous theory frequently yields practically less accurate results than the approximate models based on the physical intuition. The LPF Brandt – Lundqvist model is one such models and will be widely used in calculations below.

3. Static and polarization radiative channels in collision of charged particles with atoms and plasmas

3.1. General consideration for bremsstrahlung of fast particles on atom

A consistent quantum-electrodynamic treatment of PBR of a relativistic incident particle (IP) on a multielectron atom is complicated by the need to include into the relativistic formalism the interaction between atomic electrons, and also by the problem of summation over negative energy states in the many-electron system. At the same time, the calculation can be essentially simplified for non-relativistic atomic electrons if a non-relativistic atomic Hamiltonian is used from the very beginning and the incident particle is substituted by electromagnetic field it produces.

Let us ground the possibility of such a substitution from general principles of quantum electrodynamics. Let the field operator of the free IP $\hat{\varphi}(x)$ ($x = \{t, \mathbf{r}\}$) satisfy the Dirac equation (in units $\hbar = c = 1$):

$$(\gamma p - m_0)\,\hat{\varphi}(x) = 0\,. \tag{3.1}$$

We shall assume that the operator of an electron – positron field of atomic electrons $\hat{\psi}(x)$ satisfies the Dirac equation with the interaction:

$$\left[\gamma\left(p + e\,A^{\text{ext}}(x) + e\,\hat{A}^{\text{ae}}\right) - m\right]\hat{\psi}(x) = 0 \tag{3.2}$$

where $A^{\text{ext}}(x)$ is the potential of the external field of the nucleus, $\hat{A}^{\text{ae}}(x)$ is the operator of the electromagnetic field produced by atomic electrons that satisfies the Maxwell equation:

$$\partial^\nu \partial_\mu \hat{A}^{\text{ae}\,\mu}(x) - \partial^\mu \partial_\mu \hat{A}^{\text{ae}\,\nu}(x) = 4\,\pi\,e\,\hat{j}^\nu(x) \tag{3.3}$$

where $\hat{j}^\nu(x) = \hat{\bar{\psi}}(x)\,\gamma^\nu\,\hat{\psi}(x)$ is the operator of the atomic electron current.

Thus the interaction between atomic electrons is assumed to be taken into account in $\hat{\psi}(x)$. The state vector of a system of fields (atomic electrons, incident particle, electromagnetic field) can be represented as the product: $|\Phi_j\rangle = |j\rangle|\varphi_j\rangle|n_{\mathbf{k}\sigma}\rangle$, where $|j\rangle$ is the vector of state of interacting atomic electrons, $|\varphi_j\rangle$ the vector of state of the free incident particle, $|n_{\mathbf{k}\sigma}\rangle$ the vector of state of the electromagnetic field. The equation for $|\Phi\rangle$ in the interaction representation takes into account the interaction of currents produced by the incident and atomic particles with electromagnetic field:

$$i\partial|\Phi\rangle/\partial t = \int dr\left[e_0\hat{J}^\nu(x) - e\hat{j}^\nu(x)\right]\hat{A}_\nu(x)\,|\Phi\rangle\,,$$

$$\hat{J}^\nu(x) = \hat{\bar{\varphi}}(x)\,\gamma^\nu\,\hat{\varphi}(x)\,.$$

Hence the scattering matrix $\hat{S}$ can be written in the standard form:

$$\hat{S} = T\exp\left\{-i\int dx\,\hat{A}_\nu(x)\left[e_0\,\hat{J}^\nu(x) - e\,\hat{j}^\nu(x)\right]\right\} \tag{3.4}$$

where T stands for the chronological ordering.

The PBR amplitude that includes interaction between incident and atomic particles both between themselves and with a electromagnetic field, in the lowest order of the perturbation theory is described by the third term in expansion of the scattering operator $\hat{S}$ ($x_i \equiv i$):

$$\hat{S}_3 = (-i)^3 \, e^2 \, e_0 \int d1 \, d2 \, d3 \, T\left\{\hat{A}_\nu(1)\hat{j}^\nu(1) \, \hat{A}_\mu(2)\hat{j}^\mu(2) \, \hat{A}_\lambda(3)\hat{J}^\lambda(3)\right\}.$$
(3.5)

In obtaining this formula we have reduced the similar terms arisen due to permutation of the integration variables. Below we shall assume for simplicity that there is no exchange between the incident particle and atomic electrons. Using the commutative property of the corresponding operators expression (3.5) can be recasted using the photon field motion operator (propagator):

$$\hat{S}_3 = (-i)^2 \int d1 d2 \, \hat{A}_\nu(1) \, T\left\{e^2 \hat{j}^\nu(1) \, \hat{j}^\mu(2)\right\} \int d3 \, e_0 \, D_{\mu\lambda}(2,3) \, \hat{J}^\lambda(3)$$
(3.6)

where $D_{\mu\lambda}(2,3) = iT\langle 0|\hat{A}_\mu(2) \, \hat{A}_\lambda(3)|0\rangle$ means the photon propagator.

Formula (3.6) still keeps one non-paired $\hat{A}$-operator, which corresponds to changing the electromagnetic field by one photon.

Calculating matrix element $\hat{S}_3$ using the initial and final states of the system, we arrive at:

$$S_{3,\text{fi}}^{\text{pol}} = (-i)^2 \int d1 \, d2 \, A_{k\sigma\nu}^*(1) \, L_{\text{fi}}^{\nu\mu}(1,2) \, A_{\mu,\text{fi}}^{(0)}(2).$$
(3.7)

Here the following operators are introduced:

$$L_{\text{fi}}^{\nu\mu}(1,2) = e^2 \, \langle f|T\left\{\hat{j}^\nu(1) \, \hat{j}^\mu(2)\right\}|i\rangle$$
(3.8)

— the relativistic analog of the electromagnetic field scattering tensor by atom, and

$$A_{\mu,\text{fi}}^{(0)}(2) = -e_0 \int d3 D_{\mu\nu}(2,3) \, \langle\varphi_\text{f}|\hat{J}^\nu(3)|\varphi_\text{i}\rangle$$
(3.9)

— the 4-potential of the virtual photon produced by the incident particle in the scattering process $|\varphi_\text{i}\rangle \rightarrow |\varphi_\text{f}\rangle$. Note that the virtual photon potential $A_\text{fi}^{(0)}$ could be obtained from Maxwell equations (3.3) by substituting the matrix element of the IP transition current $\langle\varphi_\text{f}| \, \hat{J}^\mu(3) \, |\varphi_\text{i}\rangle$ on the right-hand side.

Formula (3.7) for the PBR amplitude allows for interpretation in terms of the amplitude of virtual photon $A_{\text{fi}}^{(0)}$ scattering (conversion) into the real photon on atomic electrons.

It is easy to demonstrate that the same expression for the PBR amplitude can be derived from another form of the interaction Hamiltonian:

$$V' = -e \int d\mathbf{r} \left\{ \hat{A}_\nu(x) + A_{\text{fi},\nu}^{(0)}(x) \right\} \hat{j}^\nu(x). \tag{3.10}$$

Here the incident particle is substituted by the electromagnetic field $A_{\text{fi}}^{(0)}$ it generates and thus is excluded from consideration as additional dynamic degree of freedom. The field $A_{\text{fi}}^{(0)}$ can be considered known from Eqn. (3.9); this is the so-called defined current approximation [38]. In this approximation the PBR amplitude can be calculated in the standard way in the 2nd order of the perturbation theory. After calculating the corresponding matrix element we obtain:

$$S_{2,\text{fi}}'^{\text{pol}} = (-i)^2 \int d1 \, d2 \, A_{\mathbf{k}\sigma,\nu}^*(1) \langle f | T \left\{ e^2 \, \hat{j}^\nu(1) \, \hat{j}^\mu(2) \right\} | i \rangle \, A_{\text{fi}}^{(0)}(2). \tag{3.11}$$

Comparison of Eqn. (3.7) with (3.11) yields

$$S_{3,\text{fi}}^{\text{pol}} = S_{2,\text{fi}}'^{\text{pol}}.$$

Thus the PBR amplitude can be calculated by substituting the incident particle by the field it produces using Eqn. (3.9). Then in the considered here case of non-relativistic atomic electrons the only relativistic degree of freedom, the incident particle, will be excluded by such a substitution, and non-relativistic formalism for the BR amplitude computation can be applied.

Note that the substitution of a particle by its field is widely used also for the BR calculation in the Bethe–Heitler approximation by the equivalent photon method, when the atom's field in the IP rest frame is substituted by some equivalent photons which are Compton-scattered on the incident particle into the bremsstrahlung photons.

Let us substitute the IP by its field to calculate the PBR amplitude for a non-relativistic multielectron atom ($Z \ll 137$), ignoring the exchange between the incident and bound electrons. We use the axial gauge for the electromagnetic potential ($A_0 = 0$). The non-relativistic Hamiltonian of the atomic electron perturbation by electromagnetic field has the form:

$$\hat{V} = \frac{e}{2\,m} \sum_j \left\{ \hat{\mathbf{p}}_j \hat{\mathbf{A}}(\mathbf{r}_j, t) + \hat{\mathbf{A}}(\mathbf{r}_j, t)\,\hat{\mathbf{p}}_j + e\,\hat{\mathbf{A}}^2(\mathbf{r}_j, t) \right\} \qquad (3.12)$$

where $\hat{\mathbf{p}}_j = -i\,\nabla_j$, $\hat{\mathbf{A}} = \hat{\mathbf{A}}^{\mathrm{ph}} + \mathbf{A}_{\mathrm{fi}}^{(0)}$ is the total vector – potential, the operator $\hat{\mathbf{A}}^{\mathrm{ph}}$ describes the photon field ($kx = \omega t - \mathbf{kr}$, $\omega = |\mathbf{k}|$)

$$\hat{\mathbf{A}}^{\mathrm{ph}}(x) = \sum_{\mathbf{k},\sigma} \sqrt{\frac{2\,\pi}{\omega}} \left\{ \mathbf{e}_{\mathbf{k},\sigma}\,\hat{c}_{\mathbf{k},\sigma}\,\exp(-ikx) + \mathbf{e}^*_{\mathbf{k},\sigma}\,\hat{c}^+_{\mathbf{k},\sigma}\,\exp(ikx) \right\}.$$

$$(3.13)$$

Here $\mathbf{e}_{\mathbf{k},\sigma}$ is the photon polarization unitary vector, $c^+_{\mathbf{k},\sigma}, c_{\mathbf{k},\sigma}$ are the creation and annihilation operators. $\mathbf{A}_{\mathrm{fi}}^{(0)}$ is given by formula (3.9) and is the external field produced by the incident particle.

Passing to the interaction representation

$$\hat{V}_{\mathrm{int}} = \exp(iH_{\mathrm{a}}t)V\exp(-iH_{\mathrm{a}}t)\,,$$

the scattering operator reads:

$$\hat{S} = T \exp\left\{ -i \int\limits_{-\infty}^{\infty} \hat{V}_{\mathrm{int}}(t)\,dt \right\}. \qquad (3.14)$$

The contribution to the PBR amplitude in the lowest order of the perturbation theory (in the second order of the electron charge) is due to the first and the second terms of expansion S. The zeroth term of this expansion, unity, corresponds obviously to a steady-state of the system. In the first-order term the contribution to the process is due to the term containing the square of the total vector – potential; in the second-order term, it is due to the term containing $\hat{\mathbf{p}}\hat{\mathbf{A}} + \hat{\mathbf{A}}\hat{\mathbf{p}}$. According to the PBR physical picture, it is necessary to retain the terms including the mixed product of $\hat{\mathbf{A}}_{\mathrm{ph}}$ and $\mathbf{A}_{\mathrm{fi}}^{(0)}$. Thus the matrix element of the process takes the form:

$$S_{\mathrm{fi}}^{\mathrm{pol}} = S_{\mathrm{fi}}^{(1)} + S_{\mathrm{fi}}^{(2)}$$

where

$$
S_{\text{fi}}^{(1)} = -i \, \langle \Phi_f | \int\limits_{-\infty}^{\infty} dt \, \exp(iH_a t) \, \frac{e^2}{2m}
$$

$$
\times \sum_{j=1}^{N} 2 \, \hat{A}^{\text{ph}}(\mathbf{r}_j, t) \, \mathbf{A}_{\text{fi}}^{(0)}(\mathbf{r}_j, t) \, \exp(-iH_a t) \, |\Phi_i\rangle \, . \quad (3.15)
$$

Here $|\Phi_j\rangle = |j\rangle \, |n_{\mathbf{k},\sigma}\rangle$, because the variables related the incident particle have already been accounted for in $\mathbf{A}_{\text{fi}}^{(0)}$. Thus from Eqn. (3.15) we obtain

$$
S_{\text{fi}}^{(1)} = -2i\pi \, \delta \left(\varepsilon_{\text{f}} + E_{\text{f}} + \omega - \varepsilon_{\text{i}} - E_{\text{i}} \right)
$$

$$
\times \sqrt{\frac{2\pi}{\omega}} \, \mathbf{e}_{\mathbf{k},\sigma}^{*} \, \mathbf{A}_{\text{fi}}^{(0)}(q_1) \, \langle f | \sum_{j=1}^{N} \exp\left(-i\mathbf{q}\mathbf{r}_j\right) |i\rangle \, \frac{e^2}{m} \quad (3.15a)
$$

where $\mathbf{A}_{\text{fi}}^{(0)}(q_1)$ is the spatial-time Fourier-image of the incident particle field calculated on the four-vector $q_1 = \{\varepsilon_{\text{f}} - \varepsilon_{\text{i}}, \, \mathbf{p}_{\text{f}} - \mathbf{p}_{\text{i}}\}$. We neglect spin effects. For $S_{\text{fi}}^{(2)}$ we obtain the analogical expression:

$$
S_{\text{fi}}^{(2)} = -\frac{1}{2} \, \langle \Phi_f | T \iint dt \, dt' \, \hat{V}_{\text{int}}(t) \, \hat{V}_{\text{int}}(t') |\Phi_i\rangle \, . \quad (3.16)
$$

After simple transformations $S_{\text{fi}}^{(2)}$ is reduced to the form:

$$
S_{\text{fi}}^{2} = -e^2 \, 2\pi \, \delta(\Delta E_{\text{i}}) \sqrt{\frac{2\pi}{\omega}} \, \mathbf{e}_{\mathbf{k},\sigma,l}^{*} \, A_{\text{fi},s}^{(0)}(q_1)
$$

$$
\times \langle f | \int d\tau \, \exp(i\omega\tau) \, \hat{j}^{\, l}(\mathbf{k}, \tau) \, \hat{j}^{\, s}(\mathbf{q}_1) |i\rangle \quad (3.16a)
$$

where

$$
j^{\, l}(\mathbf{k}, \tau) = \exp(iH_a \tau)
$$

$$
\times \frac{1}{2m} \sum_{j=1}^{N} \left\{ \hat{p}_j^{\, l} \, \exp(-i\mathbf{k}\mathbf{r}_j) + \exp(-i\mathbf{k}\mathbf{r}_j) \, \hat{p}_j^{\, l} \right\} \exp(-iH_a \tau)
$$

is the spatial Fourier-image of the atomic electron current operator in the interaction representation.

Summing $S_{\text{fi}}^{(1)}$ and $S_{\text{fi}}^{(2)}$, we obtain the PBR amplitude in the form:

$$S_{\text{fi}}^{\text{pol}} = 2\pi\, i\, \delta\left(\varepsilon_{\text{f}} + E_{\text{f}} + \omega - \varepsilon_{\text{i}} - E_{\text{i}}\right)\left(q_1^0\right)^2$$

$$\times \sqrt{\frac{2\pi}{\omega}}\, e_{\mathbf{k},\sigma,l}^*\, A_{\text{fi},s}^{(0)}(q_1)\,\langle f|\hat{c}^{ls}(k,\mathbf{q}_1)|i\rangle, \qquad (3.17)$$

$$q_1^0 = \varepsilon_{\text{f}} - \varepsilon_{\text{i}}.$$

In equation (3.17) $\hat{c}^{ls}(k,\mathbf{q}_1)$ is the operator of electromagnetic field scattering on atom in the non-relativistic (by atomic electrons) approximation, which can be recasted to the form:

$$\hat{c}^{ls}(k,\mathbf{q}_1) = \frac{e^2}{m\,(q_1^0)^2}$$

$$\times \left[im\int_{-\infty}^{\infty} d\tau\, \exp(i\omega\tau)\, T\left\{\hat{j}^{\,l}(\mathbf{k},\tau)\,\hat{j}^s(\mathbf{q}_1,0)\right\} - \delta^{ls}\,\hat{n}(\mathbf{q})\right]. \quad (3.18)$$

Here $\hat{n}(\mathbf{q}) = \sum\limits_{j=1}^{N} \exp(-i\,\mathbf{q}\mathbf{r}_j)$ is the Fourier image of the atomic electron density operator.

Formula (3.17) for the PBR amplitude corresponds to the PBR interpretation as being the incident particle proper field scattering on the atomic electrons into the bremsstrahlung photon.

Analysis of the initial relativistic expression from which Eqn. (3.18) follows confirms that the first term in the squared brackets in (3.18) comes from the sum over the positive part of the atomic electron spectrum and describes the electromagnetic field scattering on the atomic electron current. The second term appears after reducing the sum over negative energy spectrum states and describes the field scattering on the atomic electron charge.

Let us write the matrix element $c_{\text{fi}}^{ls}(k,\mathbf{q}_1)$ through the sum over intermediate states of atomic electrons:

$$c_{\text{fi}}^{ls}(k,\mathbf{q}_1) = \frac{e^2}{m(q_1^0)^2}$$

$$\times \left\{m\sum_n \left[\frac{j_{fn}^l(\mathbf{k})j_{ni}^s(\mathbf{q}_1)}{\omega_{fn} + \omega + i0} + \frac{j_{fn}^s(\mathbf{q}_1)j_{ni}^l(\mathbf{k})}{\omega_{in} - \omega + i0}\right] - \delta^{ls}\,n_{fi}(\mathbf{q})\right\}. \quad (3.19)$$

In the particular case of a spherically-symmetric state $|i\rangle$ for $f = i$, $\mathbf{k} = \mathbf{q}_1 = 0$, Eqn. (3.19) gives the well-known expression for atom's dipole polarizability:

$$c_{ii}^{ls}\left(\mathbf{q}_1, \mathbf{k} \to 0\right) \to \alpha(\omega)\,\delta^{ls} = \delta^{ls}\,\frac{e^2}{m}\sum_n \frac{f_{in}}{\omega_{in}^2 - \omega^2}. \qquad (3.20)$$

Here f_{in} is the oscillator strength for the transition $i \to n$. Equations (3.17) – (3.20) suppose the detuning off the resonance Δ be sufficiently large so that $\Delta = \left|\omega - \omega_{f(i)n}\right| \gg \Gamma_{f(i)n}$, where $\Gamma_{f(i)n}$ is the width of the transition $n \to f(i)$. Otherwise the line width of the corresponding transitions should be properly accounted for in these expressions.

Now let us calculate the amplitude of the static ("traditional") bremsstrahlung radiation produced by photon emission of the incident particle, including the possible excitation of atomic electrons. Here again we use the bremsstrahlung radiation interpretation in terms of scattering of virtual photons into real. Now virtual photons are generated by atom (nucleus and bound electrons). Atoms at rest and non-relativistic atomic electrons mainly produce longitudinal virtual photons. In this case it is convenient to use the Coulomb gauge of the electromagnetic potential (div $\mathbf{A} = 0$), since then only its time component can be considered. Spatial components in the Coulomb gauge describe the transversal part of the field and are small in the case under study. The time component of the virtual photon potential produced by an atom, according to Eqn. (3.9), reads

$$A_{\mathrm{fi}}^0 = -\int d1'\, D_{00}(1, 1')\, \langle f|\hat{J}^0(1')|i\rangle \qquad (3.21)$$

where $\hat{J}^0(1) = Ze\,\delta(\mathbf{r}_1 - \mathbf{r}_0) - e\sum_{j=1}^{N}\delta(\mathbf{r}_1 - \mathbf{r}_j)$ is the atomic charge density operator in the coordinate representation, $\mathbf{r}_0$ is the radius-vector of the nucleus. Applying standard rules of quantum electrodynamics it is easy to obtain the static bremsstrahlung radiation amplitude:

$$S_{\mathrm{fi}}^{\mathrm{st}} = -2\pi i\sqrt{\frac{2\pi}{\omega}}\,e_0^2\,e_{\mathbf{k},\sigma,\nu}^*\,T^\nu(p_{\mathrm{f,i}}; k)A_{\mathrm{fi}}^0(q)\delta(\varepsilon_{\mathrm{f}} + E_{\mathrm{f}} + \omega - \varepsilon_{\mathrm{i}} - E_{\mathrm{i}}). \qquad (3.22)$$

Here the following notations are introduced:

$$T^\nu = \frac{\bar{u}_f}{\sqrt{2\varepsilon_f}} \left\{ \gamma^\nu \frac{p_f \gamma + \gamma k + m_0}{(p_f + k)^2 - m_0^2} \gamma^0 + \gamma^0 \frac{p_i \gamma - \gamma k + m_0}{(p_i - k)^2 - m_0^2} \gamma^\nu \right\} \frac{\bar{u}_i}{\sqrt{2\varepsilon_i}},$$

(3.23)

$$A_{fi}^0(\mathbf{q}) = \left(4\pi / \mathbf{q}^2 \right) \left\{ \delta_{fi} Ze - e\, n_{fi}(\mathbf{q}) \right\}.$$

(3.24)

Thus the total BR amplitude of a relativistic incident particle on a non-relativistic atom ($Z \ll 137$) including polarization mechanism and the possible excitation of atomic electrons has the form

$$S_{fi}^{Br} = S_{fi}^{st} + S_{fi}^{pol}$$

(3.25)

where S_{fi}^{pol} and S_{fi}^{st} are given by Eqns (3.17) and (3.22), respectively.

Using expressions obtained for amplitudes, we write the BR spectral cross section in the form

$$\frac{d\sigma^{Br}(\omega)}{d\omega} = \frac{\varepsilon_i}{|\mathbf{p}_i|} \sum_{f,\sigma} \frac{d\Omega_\mathbf{k}}{(2\pi)^3} \frac{d\mathbf{q}}{(2\pi)^3} \lim_{T\to\infty} \frac{\left| S_{fi}^{Br}(\sigma; \mathbf{p}_{f,i}; \mathbf{k}) \right|^2}{T}.$$

(3.26)

Here $d\Omega_\mathbf{k}$ is the elementary solid angle along the photon wave vector direction $\mathbf{k}$, T is the normalization parameter which has the sense of the time, the summation is made over the emitted photon polarizations (σ) and final states of the atom (f). Below we shall assume a Born incident particle with non-degenerate initial state.

Bearing in mind the explicit form of S_{fi}^{Br}, Eqn. (3.26) can be rewritten in the form:

$$\frac{d\sigma^{Br}(\omega)}{d\omega} = \frac{\varepsilon_i}{|\mathbf{p}_i|} \sum_{f,\sigma} \omega^2 \frac{d\Omega_\mathbf{k}}{(2\pi)^3} \frac{d\mathbf{q}}{(2\pi)^3} 2\pi\delta(\Delta E) \frac{2\pi}{\omega}$$

$$\times \left| e_{\mathbf{k}\sigma,l}^* \left\{ e_0^2 T^l \frac{4\pi}{q^2} (Ze\delta_{fi} - en_{fi}(\mathbf{q})) + (q_1^0)^2 c_{fi}^{ls} A_{fi,s}^0 \right\} \right|^2$$

(3.27)

that includes three terms:

$$\frac{d\sigma^{Br}(\omega)}{d\omega} = \frac{d\sigma^{st}}{d\omega} + \frac{d\sigma^{pol}}{d\omega} + \frac{d\sigma^{int}}{d\omega}.$$

(3.28)

The last term in Eqn. (3.28) describes interference between static and polarization terms in BR; T^l and c_{fi}^{ls} are given by Eqns (3.23) and (3.19), correspondingly.

Below we shall assume the IP motion to be slightly perturbed by bremsstrahlung radiation, i.e. that $|\mathbf{q}_1| \ll |\mathbf{p}_{f,i}|$. Then A_{fi}^0 reads $(c = 1)$:

$$\mathbf{A}^{(0)}(q) = \frac{4\pi e_0}{q^0} \frac{\mathbf{v}\, q^0 - \mathbf{q}}{(q^0)^2 - \mathbf{q}^2}, \qquad q^0 = \mathbf{q}\mathbf{v}_0. \tag{3.29}$$

Here $\mathbf{v}_0$ is the incident particle velocity. In the same approximation, the function $\mathbf{T}$ (see definition (3.23)) reads:

$$\mathbf{T} = \frac{\mathbf{q}_1}{m_0\, \gamma\, (\omega - \mathbf{k}\mathbf{v}_0)}, \qquad \gamma = \varepsilon_i/m_0. \tag{3.30}$$

Expression (3.23) is the most general form for the bremsstrahlung radiation cross section on an atom. Neglecting internal degrees of freedom of IP and atomic nucleus, it describes the atomic electron contribution to the BR process. The static BR cross section can be found from Eqn. (3.23) after simple transformations:

$$\frac{d\sigma^{st}}{d\omega} = \frac{\omega}{v_0} \int \frac{d\Omega_{\mathbf{k}}\, d\mathbf{q}}{(2\pi)^4} \int dt\, e^{it(\omega + q_1^0)}$$

$$\times \sum_\sigma \left| \mathbf{e}_{\mathbf{k},\sigma}^* \mathbf{T} \right|^2 \frac{e_0^4\, e^2}{\mathbf{q}^2} \langle i|(Z - \hat{n}(-\mathbf{q}))\, (Z - \hat{n}(\mathbf{q}, t))|i\rangle. \tag{3.31}$$

If the excitation energy of atomic electrons can be disregarded compared to the emitted photon frequency ω, in Eqn. (3.31) we can put $\hat{n}(\mathbf{q}, t) \approx \hat{n}(\mathbf{q}, 0)$ to obtain:

$$\frac{d\sigma^{st}}{d\omega} = \frac{\omega}{v_0} \int \frac{d\Omega_{\mathbf{k}}\, d\mathbf{q}}{(2\pi)^3}\, \delta(q_1^0 + \omega)\, [\mathbf{n}\,\mathbf{T}]^2\, \frac{e_0^4\, e^2}{\mathbf{q}^2} \langle i||Z - \hat{n}(\mathbf{q})|^2|i\rangle,$$

$$\mathbf{n} = \frac{\mathbf{k}}{k}. \tag{3.32}$$

In deriving Eqn. (3.32) we have used the equality $\sum_\sigma e_{\mathbf{k}\sigma,l}^* e_{\mathbf{k}\sigma,s} = \delta_{ls} - n_l\, n_s$.

Equation (3.32) coincides with the result obtained by Lamb and Wheeler [39], who were the first to study consistently the contribution to the static BR due to atomic electrons.

In the case of a heavy $(m_0 \gg m)$ IP, the first term under the modulus sign in Eqn. (3.27) can be neglected compared to the second term, because $|\mathbf{T}| \propto 1/m_0$, while $\mathbf{A}^{(0)}(\mathbf{q})$ and $\hat{c}^{ls}(k, \mathbf{q}_1)$ are independent of mass. Then the total BR cross section on an atom is reduced to the PBR cross section, for which from Eqn. (3.26) we find:

$$\frac{d\sigma^{\text{pol}}}{d\omega} = \frac{\omega}{v_0} \int \frac{d\Omega_\mathbf{k}\, d\mathbf{q}}{(2\pi)^5} \left(\delta_{ls} - n_l\, n_s\right) (q_1^0)^4 A_{\text{fi},s'}^0(q_1)\, A_{\text{fi},l'}^0(q_1)$$

$$\times \int dt\, e^{iq^0 t} \langle i|\hat{c}^{\,sl'*}(0)\, \hat{c}^{\,ls'}(t)|i\rangle . \tag{3.33}$$

Here we put

$$\hat{c}^{\,ls}(t) = \exp(i\, H_a\, t)\, \hat{c}^{\,ls}(0)\, \exp(-i\, H_a\, t) . \tag{3.34}$$

Consider now PBR without atomic excitation (elastic PBR). Its cross section is given by the addend with $f = i$ in the second term under the modulus sign in Eqn. (3.27):

$$\frac{d\sigma_{ii}^{\text{pol}}}{d\omega} = \frac{\omega}{v_0} \int \frac{d\Omega_\mathbf{k}\, d\mathbf{q}}{(2\pi)^4} \left(\delta_{ls} - n_l\, n_s\right) (q_1^0)^4 A_h^{(0)}(q_1)\, A_r^{(0)}(q_1)$$

$$\times \delta(q_1^0 + \omega)\, \langle i|\hat{c}^{\,lh}|i\rangle\langle i|\hat{c}^{\,sr*}|i\rangle . \tag{3.35}$$

Inside the frequency range $\omega \ll p_a v_0$ ($p_a \approx Z^{1/3}\, m\, e^2$ is the characteristic atomic momentum) the main contribution to the process is due to moduli $|\mathbf{q}_1| \ll p_a$ permitted by energy conservation. In the opposite case (when $|\mathbf{q}_1| \gg p_a$) PBR with excitation and ionization of atom should prevail. So in this case the dipole approximation for the scattering tensor can be applied:

$$c_{ii}^{lh}(k, \mathbf{q}_1) \to \delta^{lh}\, \alpha_i(\omega)\, \theta\, (p_a - |\mathbf{q}_1|) \tag{3.36}$$

and instead of Eqn. (3.35) we obtain

$$\frac{d\sigma_{ii}^{\text{pol}}}{d\omega} \approx \frac{\omega}{v_0} \int \frac{d\Omega_\mathbf{k}\, d\mathbf{q}}{(2\pi)^4} \left[\mathbf{n}\, \mathbf{A}^{(0)}(q_1)\right]^2 \delta(q^0)\, \theta\, (p_a - |\mathbf{q}_1|) \left|\omega^2\, \alpha_i(\omega)\right|^2,$$

$$\omega < p_a v_0 . \tag{3.37}$$

The approximation used corresponds to the Born–Bethe approximation in theory of atomic excitation by electronic collision.

Equation (3.37) yields frequency-angular distribution of elastic PBR in the given frequency range

$$\frac{d\sigma_{ii}^{\text{pol}}(\omega, \vartheta)}{d\omega} = \frac{2\,e_0^2}{v_0^2}\frac{d\omega}{\omega}\left|\omega^2\,\alpha_i(\omega)\right|^2(1 + \cos^2\vartheta)\sin\vartheta\,d\vartheta\,\ln\left(\frac{\gamma\,p_a\,v_0}{\omega}\right).$$

(3.38)

In the derivation of Eqn. (3.38) we have neglected terms of the order of unity compared to the large logarithm.

Equation (3.38) leads to consequences: (a) in contrast to static BR, polarization BR of an ultra-relativistic IP ($\gamma \gg 1$) at frequencies $\omega < p_a v_0$ is not narrow-beamed and has a dipole-like angular distribution; and (b) the PBR cross section increases logarithmically with the IP energy increase in the ultra-relativistic limit at $\omega < p_a v_0$.

These characteristic features of the relativistic IP PBR allows a simple physical interpretation. The logarithmic PBR cross section growth with the IP energy is due to peculiarities of the proper electromagnetic field of the relativistic charged particle. The spatial distribution of the potential of this field at the frequency ω is given by the formula

$$A^{(0)}(\omega) \propto \exp\left(i\frac{\omega}{v_0}(z - v_0\,t) - i\frac{\omega\,\rho}{\gamma\,v_0}\right).$$

(3.39)

Here z, ρ are cylindric coordinates of the IP field. Equation (3.39) suggests that the transversal size of the field is of the order of $\rho_{\max} \approx \gamma\,v_0/\omega$, and correspondingly the minimum transversal transmitted momentum is $|\mathbf{q}_\perp|_{\min} \approx \omega/\gamma\,v_0$. Hence the spectral PBR cross section (in the Born approximation) $d\sigma^{\text{pol}}(\omega) \propto \ln\left(|\mathbf{q}_\perp|_{\max}/|\mathbf{q}_\perp|_{\min}\right)$ entails the second PBR feature noted above. In the case of static BR on neutral atoms, the maximum size of the field scattering on the IP into a bremsstrahlung photon is determined by the atomic size.

For elastic PBR inside the frequency range $I \ll \omega \ll m$ (I is the atom's ionization potential), the high frequency asymptotic for the scattering operator can be used:

$$\hat{c}^{ls}(k, \mathbf{q}_1) \approx -\frac{e^2}{m\,(q_1^0)^2}\,\hat{n}(\mathbf{q})\left\{\delta^{ls} + \frac{q_1^l\,q_1^s}{2\,m\,\omega}\right\}, \quad I \ll \omega \ll m. \quad (3.40)$$

Formula (3.40) is obtained by expanding the matrix element c_{fi}^{ls} (3.19) in a power series of the ratio $|\omega_{jn}|/\omega$ ($j = f, i$). Here the terms in the sum over intermediate states with $|w_{jn}| > \omega$ provide minor contribution to c_{fi}^{ls} for $\omega \gg I$. Substituting Eqn. (3.40) into (3.33) we find

$$\frac{d\sigma_{ii}^{\text{pol}}}{d\omega} = \frac{\omega}{v_0} \int \frac{d\Omega_{\mathbf{k}}\, d\mathbf{q}}{(2\pi)^4}\, \delta(q^0) \left(\frac{e^2}{m}\right)^2 |n_{ii}(\mathbf{q})|^2$$

$$\times \left[\mathbf{n}, \left(\mathbf{A}^{(0)}(\mathbf{q}) + \frac{\mathbf{q}_1 \left(\mathbf{q}_1 \mathbf{A}^{(0)}(\mathbf{q}_1)\right)}{2\,m\,\omega}\right)\right]^2,$$

$$I \ll \omega \ll m\,. \tag{3.41}$$

The quantity $n_{ii}(\mathbf{q})$ is the (static) form factor of the atomic core in the state $|i\rangle$.

Now we calculate the cross section for PBR with excitation (including ionization) of atom for $m \gg \omega \gg I$. Substituting into Eqn. (3.33) expression for $\hat{c}^{lh}$ (3.40), we obtain

$$\frac{d\sigma_{ii}^{\text{pol}}}{d\omega} = \frac{\omega}{v_0} \int \frac{d\Omega_k\, dq}{(2\pi)^4} \left(\frac{e^2}{m}\right)^2$$

$$\times \left[n, \left(A^{(0)}(q_1) + \frac{q_1(q_1\, A^{(0)}(q_1))}{2m\omega}\right)\right]^2 S_{ii}(q)\,. \tag{3.42}$$

Here we introduced the quantity

$$S_{ii}(q) = \frac{1}{2\pi} \int_{-\infty}^{\infty} dt\, \exp(iq^0 t)\, \langle i|\hat{n}(-q)\, \hat{n}(q,t)|i\rangle\,, \tag{3.43}$$

that we shall call the dynamic form factor (DFF) in accordance with the terminology accepted for description of effects in a medium.

The simplest analytical approximation for $S_{ii}(q)$ has the form

$$S_{ii}(q) \approx \theta\left(|q| - p_{\mathrm{a}}\right) \delta\left(q^0 + \frac{q_1^2}{2m}\right) N + \theta\left(p_{\mathrm{a}} - |q|\right) \delta(q^0)\, N^2\,. \tag{3.44}$$

Combining Eqns (3.42) and (3.44), the PBR spectral cross section in the approximation considered can be derived in the form

$$\frac{d\sigma_{ii}^{\text{pol}}}{d\omega} = \frac{16e_0^2 e^4}{3m^2 v_0^2} \left\{ \theta(p_a v_0 - \omega) \left[N^2 \ln\left(\frac{\gamma p_a v_0}{\omega}\right) + N \ln\left(\frac{m_0 v_0}{p_a}\right) \right] \right.$$

$$\left. + \theta(\omega - p_a v_0) N \ln\left(\frac{\gamma m_0 v_0^2}{\omega}\right) \right\}. \tag{3.45}$$

Note that the total PBR cross section (including excitation and ionization of atom) (3.45) allows the correct limiting transition to the case $Z = 0$, which corresponds to setting $p_a = 0$ in Eqn. (3.45). Then the term in the square brackets describing "elastic" PBR disappears and the remaining last term in the curly brackets describes radiation of a slow free recoil electron in collision with a relativistic charged particle, which complies with the physical picture of the phenomenon.

3.2. The dynamic form factor formalism in description of radiation from fast particles in a plasma

In order to calculate BR generated by a fast particle on an ion in plasma we shall use the problem setting characteristic for PBR on atom (see Sect. 3.1). Consider emission of a transversal photon by medium electrons, both free and bound, during a rapid IP inelastic scattering. The virtual photon scattering into real ones on the polarization charge around IP will be neglected. This is correct for a non-relativistic plasma at frequencies $\omega \gg \omega_{\text{pe}}$, $v_0 \gg v_{Te}$ (ω_{pe} is the electron plasma density, v_{Te} is the thermal velocity of plasma electrons). Then IP plays the role of the virtual photon "source" and can be substituted, as was shown in the previous section, by the corresponding electromagnetic field (the IP initial and final states are assumed to be known) according to Eqn. (3.9). Next, the IP motion is supposed to be weakly perturbed during BR process: $|\mathbf{q}_1| = |\mathbf{p}_f - \mathbf{p}_i| \ll |\mathbf{p}_{i,f}|$ ($\mathbf{p}_{i,f}$ is the initial and final IP momenta, respectively), so the transition current density reads (the normalization coefficient is set unity)

$$J_{\text{fi}}^{\nu} = v_0^{\nu} \qquad (3.46)$$

where v_0^{ν} is the incident particle 4-velocity.

Recall that polarization affects the internal electromagnetic field of the medium, which is "coated". To account for this effect, we substitute the photon propagator in a vacuum in Eqn. (3.9) by the photon propagator in a medium, whose Fourier-image for an isotropic medium in the axial gauge ($A_0 = 0$) is given by the expression [40]:

$$D_{mn}(q) = \frac{4\pi}{(q^0)^2} \left\{ \frac{q_m q_n}{\mathbf{q}^2 \varepsilon_q^l} + \frac{(q^0)^2}{(q^0)^2 \varepsilon_q^t - \mathbf{q}^2} \left(\delta_{mn} - \frac{q_m q_n}{\mathbf{q}^2} \right) \right\} \qquad (3.47)$$

where $m, n = 1, 2, 3$; ε_q^l, ε_q^t are the longitudinal and transversal components of medium's dielectric permeability on the $q = \{q^0, \mathbf{q}\}$ 4-vector. Note that propagator (3.47) coincides, to within a factor, with the expression for the linear Green function for Maxwell's equations. Equation (3.47) for the photon propagator in the medium is linked to the medium's polarization $P_{mn}(q)$ by the relationship

$$D_{mn}^{-1}(q) = D_{mn}^{(0)\,-1}(q) - P_{mn}(q)/4\pi . \qquad (3.48)$$

Introducing the longitudinal and transversal components of dielectric permeability [40] (the summation is assumed over repeating indexes):

$$\varepsilon_q^l = 1 - q_n q_m P_{nm}(q) \Big/ \mathbf{q}^2 (q^0)^2 , \qquad (3.49)$$

$$\varepsilon_q^t = 1 - P_{mn}(q) \left(\delta_{mn} - q_m q_n \Big/ \mathbf{q}^2 \right) \Big/ 2 (q^0)^2 \qquad (3.50)$$

the polarization tensor of the medium reads

$$P_{mn}(q) = (q^0)^2 \left[\delta_{mn} - \varepsilon_q^l \frac{q_n q_m}{\mathbf{q}^2} - \varepsilon_q^t \left(\delta_{nm} - \frac{q_n q_m}{\mathbf{q}^2} \right) \right] . \qquad (3.51)$$

Equation (3.48) yields the relationship between photon propagator in the medium (3.47) and dielectric tensor components (3.49)–(3.50).

The polarization operator for non-relativistic medium particles and $|\mathbf{q}| \ll m$ can be cast in the form:

$$
P_{nm}(q) = 4\pi \left\{ \delta_{nm} \sum_{\alpha} \frac{e_\alpha^2 n_\alpha}{m_\alpha} \right.
$$
$$
\left. + \sum_s \left[\frac{\langle 0| \sum_\alpha e_\alpha \hat{\jmath}_\alpha^n(\mathbf{q})|s\rangle \langle s| \sum_{\alpha'} e_{\alpha'} \hat{\jmath}_{\alpha'}^m(\mathbf{q})|0\rangle}{\omega_{0s} + q^0 + i0} + \text{c.c.} \right] \right\} \quad (3.52)
$$

where α is the particle sort index, n_α the number density of particles α, $\hat{\jmath}_\alpha^l(\mathbf{q})$ is the Cartesian component of the spatial Fourier-image of the current density operator of particles α, $|0,s\rangle$ are many-particle wave functions of the system's ground and excited states. Equation (3.52) is valid for zero temperature $T = 0$; for $T > 0$ Eqn. (3.52) should be averaged over the initial state. If interaction between different sorts of particles can be neglected in the zero approximation, then the equality holds $P_{lm}(q) = \sum_\alpha P_{lm}^\alpha$ and the dielectric permeability tensor components due to particles α can be obtained from Eqns (3.49), (3.50), and (3.52):

$$
\varepsilon_q^{l(\alpha)} = 1 - \frac{4\pi e_\alpha^2}{m_\alpha (q^0)^2 \mathbf{q}^2} \left\{ m_\alpha \sum_s \frac{2\omega_{0s}^3 \left|n_{0s}^{(\alpha)}(\mathbf{q})\right|^2}{\omega_{0s}^2 - (q^0)^2} + \mathbf{q}^2 n_\alpha \right\}, \quad (3.53)
$$

$$
\varepsilon_q^{t(\alpha)} = 1 - \frac{2\pi e_\alpha^2}{m_\alpha (q^0)^2 \mathbf{q}^2} \left\{ m_\alpha \sum_s \frac{2\omega_{0s}^3 \left|\left[\mathbf{q}\mathbf{j}_{0s}^{(\alpha)}(\mathbf{q})\right]\right|^2}{\omega_{0s}^2 - (q^0)^2} + 2\mathbf{q}^2 n_\alpha \right\}. \quad (3.54)
$$

Here ω_{0s} is the excitation energy of the system of particles, $\hat{n} = \sum_i e^{-i\mathbf{q}\mathbf{r}_i}$ is the spatial Fourier-components of the particle density operator.

Since the quantum-mechanical averaging is assumed in Eqns (3.53), (3.54) (and not the averaging over physically infinitesimal volume), the condition $|\mathbf{q}| \ll n^{1/3}$, which is usually employed in the classical approach, is relaxed and the dielectric permeability tensor components are determined for all values of $\mathbf{q}$ at which non-relativistic treatment of the medium stays valid.

Thus the electromagnetic potential of the virtual photon generated by the incident particle in the scattering ($\mathbf{p}_i \to \mathbf{p}_f$) is

$$\mathbf{A}_{fi}^{(0)}(x) = \mathbf{A}^{(0)}(q) \exp\left[i\left(q^0 t - \mathbf{q}\,\mathbf{r}\right)\right], \quad A_m^{(0)}(q) = -e_0\, D_{mn}(q)\, \mathbf{v}_{0n}\,.$$
$$(3.55)$$

The total electromagnetic field perturbing plasma electrons in the bremsstrahlung process is the sum of the virtual photon field (3.53) and the free (transversal) photon field, which has the vector – potential in the form

$$\hat{\mathbf{A}}^{ph}(x) = \sum_{\mathbf{k},\sigma} \sqrt{\frac{2\pi}{\omega}} \left\{ \mathbf{e}_{\mathbf{k}\sigma}\, \hat{a}_{\mathbf{k}\sigma}\, e^{i\,(\mathbf{k}\mathbf{r}-\omega_k^t t)} + \mathbf{e}_{\mathbf{k}\sigma}^*\, \hat{a}_{\mathbf{k}\sigma}^+\, e^{-i\,(\mathbf{k}\mathbf{r}-\omega_k^t t)} \right\}.$$
$$(3.56)$$

Here $\sigma, \mathbf{k}$ are the polarization index and wave vector of the photon, respectively, $\mathbf{e}_{\mathbf{k}\sigma}$ is the polarization unit vector, $\hat{a}_{\mathbf{k}\sigma}^+, \hat{a}_{\mathbf{k}\sigma}$ are creation – annihilation operators for photons in the medium, and $\omega_{\mathbf{k}}^t$ is the frequency of the transversal photon with the wave vector $\mathbf{k}$ determined by the dispersion law:

$$\omega_{\mathbf{k}}^t = \frac{|\mathbf{k}|}{\sqrt{\varepsilon_k^t}}\,. \qquad (3.57)$$

Let us write down the explicit expression for ε_k^t with account for the bound electrons and ions. Neglecting interaction between plasma and bound electrons, we have

$$\varepsilon_k^t = 1 - \frac{\omega_{pe}^2}{\omega^2} + 4\, n_i \alpha_i(\omega)\,. \qquad (3.58)$$

Here $\alpha_i(\omega)$ is the dynamic polarizability of ionic cores at the frequency ω, and n_i is the ionic number density. In deriving Eqn. (3.58) we have used the condition $\omega \gg |\mathbf{k}|\, \mathbf{v}_{Te}$. The Doppler effect in accounting for the bound electron contribution to dielectric permeability (3.58) has been ignored.

The Hamiltonian of interaction between non-relativistic plasma electrons and electromagnetic field (using the axial gauge for the electromagnetic potential) has the form

$$\hat{V} = \frac{e}{2m} \sum_j \left\{ \hat{\mathbf{p}}_j \hat{\mathbf{A}}(\mathbf{r}_j, t) + \hat{\mathbf{A}}(\mathbf{r}_j, t)\, \hat{\mathbf{p}}_j + e\, \hat{\mathbf{A}}^2(\mathbf{r}_j, t) \right\},$$

(3.59)

$$\hat{\mathbf{A}} = \hat{\mathbf{A}}^{(0)} + \hat{\mathbf{A}}^{\mathrm{ph}}, \quad \hat{\mathbf{p}}_j = -i\nabla_j.$$

Here the summation is being made over both plasma and bound electrons of ionic cores.

The amplitude of the process under consideration is obtained in the same way as in Sect. 3.1, in the second order of the perturbation theory on the interaction $\hat{\mathbf{p}}\,\hat{\mathbf{A}} + \hat{\mathbf{A}}\,\hat{\mathbf{p}}$ and in the first order on the interaction $\hat{\mathbf{A}}^2$, including "cross-over" terms $\mathbf{A}^{(0)}\,\mathbf{A}^{\mathrm{ph}}$:

$$M_{\mathrm{fi}}(k, \sigma, q) = \sqrt{2\pi}\,\frac{e^2}{m}$$

$$\times \left\{ m \sum_s \left[\frac{\langle f|\, \mathbf{e}^*_{\mathbf{k}\sigma} \left(\hat{\mathbf{j}}^{pl}_{\mathbf{k}} + \hat{\mathbf{j}}^{b}_{\mathbf{k}} \right) |s\rangle\, \langle s|\left(\hat{\mathbf{j}}^{pl}_{\mathbf{q}_1} + \hat{\mathbf{j}}^{b}_{\mathbf{q}_1} \right) \mathbf{A}^{(0)}_{\mathbf{q}_1}|i\rangle }{\Omega_{fs} + \omega + i\Gamma_{fs}/2} + c.c. \right] \right.$$

$$\left. + \mathbf{e}^*_{\mathbf{k}\sigma} \mathbf{A}^{(0)}_{\mathbf{q}} \langle f|\, \hat{n}^{pl} + \hat{n}^{b}|i\rangle \right\}$$

(3.60)

where $\hat{\mathbf{j}}^{pl,b}_{\mathbf{k}}$, $\hat{n}^{pl,b}_{\mathbf{k}}$ are the spatial Fourier-images of the current density and charge operators for plasma (index pl) and bound electrons (index b) determined above; $|f, s, i\rangle$ are many-particle wave functions, Ω_{ij} are excitation energies with account for the Doppler effect, and Γ_{ij} are the line widths for transitions $i \to j$. We assume that the subsystem of bound electrons of each ions weakly interacts with plasma electrons and with electrons of neighboring ions (due to the sufficiently rarefied plasma density) so that the wave functions of bound electrons of an individual ion are determined only by the given ion's parameters. Moreover, plasma electrons are supposed to interact with ions as point-like objects.

Bearing in mind the above considerations, the plasma wave function can be presented as the product

$$\Psi_s(\mathbf{r}_l, \mathbf{R}_j, \rho^j_\alpha) = \Phi_s(\mathbf{r}_l, \mathbf{R}_j) \prod_j \psi^{(j)}_s(\rho^j_\alpha).$$

(3.61)

Here $\mathbf{r}_l$ is the radius–vector of the l-th plasma electron, $\mathbf{R}_j$ is the radius–vector of the j-th ion barycenter, ρ_α^j is the radius–vector from the α-th electron, belonging to the j-th ion, to ion's barycenter, Φ_s is the wave function of interacting plasma electrons and ions, and $\psi_s^{(j)}$ are the wave functions of the electronic subsystem of the j-the ion. Systems of functions Φ_s and $\psi_s^{(j)}$ (for each j) are orthonormalized and form a complete set. For simplicity, further we assume that the bound electron subsystem does not change its state during bremsstrahlung: $\psi_{s_i}^j = \psi_{s_f}^j = \psi_1 \exp(i\,\varphi_{1j})$ (φ_{1j} is the phase of the electron wave function of the j-th ion in the state ψ_1).

Substituting Eqn. (3.61) into (3.60), we find:

$$
M_{\mathrm{fi}}^{(1,1)}(\mathbf{k}, \sigma, \mathbf{q}) = \sqrt{2\pi} \left\{ \frac{e^2}{m} \langle \Phi_f | \hat{n}(\mathbf{q}) | \Phi_i \rangle \, \delta_{hm} \right. \tag{3.62}
$$

$$
\left. - \langle \Phi_f | \sum_j e^{-i\,\mathbf{q}_1\,\mathbf{R}_j}\, c_{hm}^{(1,1)\,j}(\mathbf{k}, \mathbf{q}_1, \omega - \mathbf{k}\mathbf{v}^j) | \Phi_i \rangle \right\} e_{\mathbf{k}\sigma,h}^* A_{\mathbf{q}_1,m}^{(0)} .
$$

Here $c_{hm}^{(1,1)\,j}(\mathbf{k}, \mathbf{q}_1, \omega - \mathbf{k}\mathbf{v}^j)$ is the diagonal matrix element of the scattering operator for an electromagnetic field on the j-th ion's electrons.

The account for Doppler-effect related terms in the energy denominators can be significant for non-relativistic plasma electrons near resonance frequencies of the bremsstrahlung photons. The Doppler effect can be neglected in the opposite case (for sufficiently large detunings of the resonance).

Ignoring the excitation energy of bound electrons in absorption of the momentum $\mathbf{k}$ compared to the photon frequency, Eqn. (3.62) can be recast in the form

$$
M_{\mathrm{fi}}^{(1,1)}(\mathbf{k}, \sigma, \mathbf{q}) = \sqrt{2\pi} \left\{ \frac{e^2}{m} n_{\mathrm{fi}}^{(e)}(\mathbf{q}) \, \delta_{hm} - c_{hm}^{(1,1),j}(\mathbf{k}, \mathbf{q}_1)\, n_{\mathrm{fi}}^{(i)}(\mathbf{q}) \right\}
$$

$$
\times\, e_{\mathbf{k}\sigma,h}^* A_{\mathbf{q}_1,m}^{(0)} \tag{3.63}
$$

where

$$n_{\mathrm{fi}}^{(e)}(\mathbf{q}) = \langle \Phi_f | \sum_l \exp(-i\mathbf{q}\mathbf{r}_l) | \Phi_i \rangle \,,$$

$$n_{\mathrm{fi}}^{(i)}(\mathbf{q}) = \langle \Phi_f | [b] \sum_j \exp(-i\mathbf{q}\mathbf{R}_j) | \Phi_i \rangle$$

are the matrix elements of the Fourier-images of the density operators for plasma electrons and ions, respectively.

Let us now calculate the differential cross section for bremsstrahlung of plasma electrons and ions, summed over all final states of the system of delocalized particles and averaged over their initial states (plasma is assumed to be in thermodynamic equilibrium):

$$d\sigma_{11}t(\mathbf{k},\mathbf{q}) = \frac{2\pi}{v_0} \sum_{f,i,\sigma} \delta\left(\varepsilon_{\mathrm{f}} - \varepsilon_{\mathrm{i}} + \omega + \mathbf{q}_1\mathbf{v}\right)$$

$$\times\, w(i) \left| M_{\mathrm{fi}}^{(1,1)}(\mathbf{k},\sigma,\mathbf{q}) \right|^2 \frac{\omega\, d\omega\, d\Omega_{\mathbf{k}}\, d\mathbf{q}}{(2\pi)^6} \quad (3.64)$$

where $w(i) = \exp\left(-\varepsilon_i/T\right) \Big/ \sum_s \exp\left(-\varepsilon_s/T\right)$, ε_s is the plasma particle energy. Substituting Eqn. (3.63) into (3.64) and summing over f, i we obtain:

$$d\sigma_{11}(\mathbf{k},\mathbf{q}) = \frac{(2\pi)^2}{v_0}\left(\delta_{hl} - n_h\, n_l\right) A_m^{(0)}(\mathbf{q}_1)\, A_n^{(0)}(\mathbf{q}_1)$$

$$\times \sum_{hlmn}^{pl}(k,q)\, \frac{\omega\, d\omega\, d\Omega_{\mathbf{k}}\, d\mathbf{q}}{(2\pi)^6} \sum_{hlmn}^{pl}(k,q)$$

$$= \left(\frac{e^2}{m}\right)^2 S^{(e)}(q)\, \delta_{hn}\, \delta_{ln} + (q^0)^4\, S^{(i)}(q)\, c_{hm}^{(1,1)*} c_{ln}^{(1,1)}$$

$$- 2\,(q^0)^2\, \frac{e^2}{m}\, \delta_{lh}\, S^{(ei)}(q)\, \mathrm{Re}(c_{hm}^{(1,1)}) \,. \quad (3.65)$$

Here $S^{(e)}(q)$, $S^{(i)}(q)$, $S^{(ei)}(q)$ are, respectively, the dynamic form factors of plasma electrons, ions, and the mixed electron–ion dynamic form factor determined by the expression:

$$S^{(e)}(q) = \frac{1}{2\pi} \int_{-\infty}^{\infty} dt \, e^{i q^0 t} \left\langle \hat{n}^{(e)}(\mathbf{q}, t) \hat{n}^{(e)}(-\mathbf{q}, 0) \right\rangle, \qquad (3.66)$$

$$S^{i}(q) = \frac{1}{2\pi} \int_{-\infty}^{\infty} dt \, e^{i q^0 t} \left\langle \hat{n}^{(i)}(\mathbf{q}, t) \hat{n}^{(i)}(-\mathbf{q}, 0) \right\rangle, \qquad (3.67)$$

$$S^{(ei)}(q) = \frac{1}{2\pi} \int_{-\infty}^{\infty} dt \, e^{i q^0 t} \left\langle \hat{n}^{(e)}(\mathbf{q}, t) \hat{n}^{(i)}(-\mathbf{q}, 0) \right\rangle \qquad (3.68)$$

Here $\hat{n}(\mathbf{q}, t) = \exp\left(i \hat{H} t\right) \hat{n}(\mathbf{q}) \exp\left(-i \hat{H} t\right)$, $\hat{H}$ is the Hamiltonian of the delocalized particles. The angular brackets mean both quantum-mechanical and statistical averaging according to the rule $\langle \ldots \rangle = Sp\left(e^{-\hat{H}/T} \ldots\right) / Sp\left(e^{-\hat{H}/T}\right)$, where the spur is taken over the complete set of functions Φ_s. Note that thermodynamic averaging is essential here since delocalized particles have a continuum spectrum.

Let us comment on Eqn. (3.65). Here we can single out three terms proportional to the electron, ion, and mixed form factors, respectively. The first term describes the process of scattering of the IP electromagnetic field, $\mathbf{A}^{(0)}(\mathbf{q})$, into a bremsstrahlung photon on plasma electrons, the second term describes the same process on plasma ions, and the third is the interference term. The dynamic form factors introduced by Eqns (3.66) – (3.68) represent Fourier-images of the spatial-temporal correlation functions density – density for different plasma components (see [41]). They characterize the efficiency of the energy – momentum absorption by fluctuating plasma via different plasma components. It is important to stress that such determined dynamic form factors describe interactions between plasma electrons and ions, for example their mutual screening, which is taken into account in the structure of the wave function $\Phi_s(\mathbf{r}_l, \mathbf{R}_j)$. Note that the dynamic form factor appears in some other problems (energy losses by fast particles in a medium, light scattering in a medium, X-ray diffraction, etc.). It is a well-studied quantity so its use in calculating the PBR cross sections seems to be quite relevant.

Calculations of the PBR cross section of a fast particle in a plasma necessitate the knowledge of the explicit dependence of dynamic form factors of the plasma components on the transmitted 4-vector q. To this end, we express them through characteristics of non-interacting particles — the dielectric permeability tensor components, using the means described in [42]. This method is based on the fact that interaction between plasma electrons and ions is weak and can be taken into account by the perturbation theory. Technically, in order to calculate the dynamic form factor the fluctuation-dissipative theorem is employed, linking the dynamic form factor with the linear response function of the plasma components on some external field. According to this theorem, we have for electrons

$$S^{(e)}(q) = \frac{\mathrm{Im}\,(F_{ee}(q))}{\pi e^2 \left(\exp\left(-q^0/T\right) - 1\right)}. \tag{3.69}$$

Here $F_{ee}(q)$ is the linear response function of the electron plasma component on a fictive external potential that acts only on the plasma electrons. This function can be expressed (see [42] for more detail) through the response function of the electron and ion components on the total potential, $\beta_{e,i}(q)$. As a result, interaction between plasma components proves to be fully accounted for and $F_{ee}(q)$ takes the form

$$F_{ee}(q) = \frac{\beta_e(q)\left[1 - \frac{4\pi}{\mathbf{q}^2}\beta_i(q)\right]}{1 - \frac{4\pi}{\mathbf{q}^2}\left(\beta_e(q) + \beta_i(q)\right)}. \tag{3.70}$$

Substituting Eqn. (3.70) into (3.69) yields the dynamic form factor in the form

$$S^{(e)}(q) = \frac{\left|\frac{1 - \frac{4\pi}{\mathbf{q}^2}\beta_i(q)}{\tilde{\varepsilon}_q^l}\right|^2 \mathrm{Im}(\beta_e(q)) + \left(\frac{4\pi}{\mathbf{q}^2}\right)^2 \left|\frac{\beta_e(q)}{\tilde{\varepsilon}_q^l}\right|^2 \mathrm{Im}(\beta_i(q))}{\pi e^2 \left(\exp\left(-q^0/T\right) - 1\right)}. \tag{3.71}$$

Here

$$\tilde{\varepsilon}_q^{l(\alpha)} = 1 - \frac{4\pi}{\mathbf{q}^2}\beta_\alpha(q), \quad \tilde{\varepsilon}_q^l = \sum_\alpha \tilde{\varepsilon}_q^{l(\alpha)} \tag{3.72}$$

where the "tilde" over the dielectric permeability sign means that the contribution from bound electrons is ignored.

The imaginary parts of the response functions on the total potential in plasma, which enter Eqn. (3.71), read [42]:

$$\text{Im}(\beta_\alpha(q)) = e_\alpha^2 \, \pi \, \left(e^{-q^0/T} - 1\right) \, n_\alpha \, \frac{\exp\left[-(q^0)^2/2\,\mathbf{q}^2\,v_{T\alpha}^2\right]}{\sqrt{2\pi}\,|\mathbf{q}|\,v_{T\alpha}} \, . \quad (3.73)$$

With account for Eqns (3.73) and (3.71), we finally arrive at:

$$S^{(e)}(q) = \frac{n_e}{\sqrt{2\pi}\,|\mathbf{q}|} \left\{ \frac{1}{v_{Te}} \left| \frac{\tilde{\varepsilon}_q^{l(i)}}{\tilde{\varepsilon}_q^l} \right|^2 \exp\left[-\frac{(q^0)^2}{2\,\mathbf{q}^2 v_{Te}^2}\right] \right.$$
$$\left. + \frac{Z_i}{v_{Ti}} \left| \frac{1 - \tilde{\varepsilon}_q^{l(e)}}{\tilde{\varepsilon}_q^l} \right|^2 \exp\left[-\frac{(q^0)^2}{2\,\mathbf{q}^2 v_{Ti}^2}\right] \right\} . \quad (3.74)$$

Analogically, for the form factor of ions we have

$$S^{(i)}(q) = \frac{n_i}{\sqrt{2\pi}\,|\mathbf{q}|} \left\{ \frac{1}{v_{Ti}} \left| \frac{\tilde{\varepsilon}_q^{l(e)}}{\tilde{\varepsilon}_q^l} \right|^2 \exp\left[-\frac{(q^0)^2}{2\,\mathbf{q}^2 v_{Ti}^2}\right] \right.$$
$$\left. + \frac{1}{Z_i v_{Te}} \left| \frac{1 - \tilde{\varepsilon}_q^{l(i)}}{\tilde{\varepsilon}_q^l} \right|^2 \exp\left[-\frac{(q^0)^2}{2\,\mathbf{q}^2 v_{Te}^2}\right] \right\} \quad (3.75)$$

and for the mixed form factor:

$$S^{(ei)}(q) = \frac{n_e}{\sqrt{2\pi}\,|\mathbf{q}|} \left\{ \frac{1}{v_{Te}\,Z_i} \left| \frac{1 - \tilde{\varepsilon}_q^{l(i)}}{\tilde{\varepsilon}_q^l} \right|^2 \exp\left[-\frac{(q^0)^2}{2\,\mathbf{q}^2 v_{Te}^2}\right] \right.$$
$$\left. + \frac{1}{v_{Ti}} \left| \frac{1 - \tilde{\varepsilon}_q^{l(e)}}{\tilde{\varepsilon}_q^l} \right|^2 \exp\left[-\frac{(q^0)^2}{2\,\mathbf{q}^2 v_{Ti}^2}\right] \right\} . \quad (3.76)$$

Now let us give a physical interpretation to expressions (3.74)–(3.76). Consider as an example the form factor for electrons $S^{(e)}(q)$. It can be represented in the convenient form

$$S^{(e)}(q) = |\delta n_e|_q^2 \left|1_{e,q}^{\text{eff}}\right|^2 + |\delta n_i|_q^2 \left|Z_{i,q}^{\text{eff}}\right|^2,$$
$$|\delta n_{e,i}|_q^2 = \frac{n_{e,i}}{\sqrt{2\pi}\,|\mathbf{q}|\,v_{Te,i}} \exp\left[\frac{-(q^0)^2}{2\,\mathbf{q}^2\,v_{Te,i}^2}\right], \quad (3.77)$$

$$1_{e,q}^{\text{eff}} = \frac{\tilde{\varepsilon}_q^{l(i)}}{\tilde{\varepsilon}_q^l}, \quad Z_{i,q}^{\text{eff}} = Z_i \frac{1 - \tilde{\varepsilon}_q^{l(e)}}{\tilde{\varepsilon}_q^l}. \tag{3.78}$$

Here $|\delta n_{e,i}|_q^2$ are squares of thermal fluctuations of electron and ion plasma components on the 4-vector q, $e\,Z_{i,q}^{\text{eff}}$ is the effective electron charge that screens the current in plasma on the same 4-vector, and $e\,1_{e,q}^{\text{eff}}$ is the effective charge of plasma electron including the screening action from other charges.

Such a representation for the electron dynamic form factor allows it to be interpreted as the sum of squares of thermal fluctuations of electron number density of two types. The first term in Eqn. (3.77) describes fluctuations due to separate electrons screened by other charges in plasma. The second term describes the electron charge that screens the ion plasma component charge fluctuations. A similar interpretation can be given to the ion dynamic form factor as well. The mixed electron – ion dynamic form factor represents the sum of charge fluctuations of each plasma components that screen charge fluctuations of the opposite sign. The zeros of the longitudinal part of the dielectric permeability in the above formulas correspond to charge fluctuations due to the appearance of collective excitations (quasi-particles) in plasma. In the last case, the momentum – energy excess induced by bremsstrahlung is transmitted to the collective excitations of the medium.

4. Plasma models for photoionization of atoms and ions

4.1. Application of dynamic polarizability models to calculations of the photoeffect on complex atoms

First we consider the application of the classical models to calculate the photoeffect cross section on multielectron atoms. The simplest approach is based on using Eqns (2.2) and (2.8) that represent the atom's dynamic polarizability as the electron density functional in the local plasma density approximation. This method was suggested in [31] to take into account many-particle effects in photoabsorption as the alternative approach to the single-particle description of the phenomenon. Its physical justification is expressed by equality (2.1): the interaction of radiation with atomic electrons is localized at a spatial point determined by the plasma resonance condition.

The photoabsorption cross section can be conveniently represented through the spectral distribution function of dipole excitations:

$$\sigma(\omega) = \frac{2\pi^2}{c} g(\omega), \quad \int g(\omega)\, d\omega = N_e. \tag{4.1}$$

The second formula in Eqn. (4.1) represents the well-known sum rule.

The spectral function $g(\omega)$ satisfies the equalities:

$$g(\omega) = \sum_n f_{in}\, \delta(\omega - \omega_n) = \frac{2}{\pi}\, \omega\, \mathrm{Im}\, \alpha(\omega). \tag{4.2}$$

Here the first equality is the spectral function definition and the second one follows from the optical theorem.

The spectral function corresponding to the dynamic polarizability in form (2.2) can be written down as a spatial integral over the radial electron density:

$$g(\omega) = \int d^3r\, n(r)\, \delta(\omega - \omega_p(r)). \tag{4.3}$$

Equality (4.3) is obtained from Eqn. (2.2) making use of the Sokhotsky formula and the optical theorem.

Note that within the frames of the statistical model of atom the photoionization and photoabsorption cross sections coincide due to the absence of bound states in the approximation employed.

The presence of the delta-function in Eqn. (4.3) allows the explicit integration and with the use of Eqn. (4.1) we find the following expression for the atomic photoionization cross section:

$$\sigma_{ph}^{B-L}(\omega) = \frac{4\pi^2\,\omega}{c}\, r_\omega^2\, \frac{n(r_\omega)}{|n'(r_\omega)|}. \tag{4.4}$$

Here r_ω is the solution of Eqn. (2.1) corresponding to the distance at which the plasma resonance occurs, the prime denotes differentiation with respect to radius.

For the electron density in the form like Eqn. (2.11), Eqn. (4.4) can be recasted in the form revealing the scaling of the photoionization cross section with respect to the reduced frequency $\nu = \omega/Z$:

$$\sigma_{\text{ph}}^{\text{B}-\text{L}}(\omega) = \sigma_{\text{ph}}^{\text{B}-\text{L}}\left(\nu = \frac{\omega}{Z}\right) = \frac{9\,\pi^4\,\nu}{32\,c}\,x_\nu^2\,\frac{f(x_\nu)}{|f'(x_\nu)|}, \qquad (4.5)$$

where x_ν is the solution to the equation

$$\nu = \sqrt{4\,\pi\,f(x)}. \qquad (4.6)$$

Equality (4.6) follows from Eqn. (2.1) with the account of Eqn. (2.11).

Based on Eqns (4.5) and (4.6), we now wish to analyze the spectral cross section of photoionization using different statistical models.

In addition to the Thomas – Fermi approximation for the normalized electron density $f(x = r/r_{\text{TF}})$ [Eqn. (2.10)], we shall consider linear distribution (2.13), statistical Lenz – Jensen model (2.12), and the exponential screening model in which

$$f_{\exp}(x) = \frac{128}{9\,\pi^3}\,e^{-2\,x}. \qquad (4.7)$$

The exponential (4.7) and especially linear (2.13) screening approximations for multielectron neutral atoms are very crude. However they can be useful in some other cases. Here we consider these approximations because on such a basis it is possible to construct simple analytical expressions for the photoionization cross section. Indeed, making use of Eqns (2.13) and (4.7), transcendental equation (4.6) can be easily solved, so it is possible to find the corresponding photoionization cross sections from Eqn. (4.5):

$$\sigma_{\text{ph}}^{\text{B}-\text{L (lin)}}(\omega = Z\nu) = \frac{27\,\pi^6\,x_0^6}{2^{14}\,c}\,\nu^3\left(1 - \frac{3\,\pi^2\,x_0^3}{2^9}\,\nu^2\right)^2,$$
$$(4.8)$$

$$\nu \leq \sqrt{\frac{512}{3\,\pi^2\,x_0^3}} \cong \frac{4.16}{x_0^{3/2}},$$

$$\sigma_{\text{ph}}^{\text{B}-\text{L (exp)}}(\omega = Z\,\nu) = \frac{9\,\pi^4}{64\,c}\,\nu\,\ln^2\left(\frac{16\,\sqrt{2}}{3\,\pi\,\nu}\right), \qquad \nu \leq \frac{16\,\sqrt{2}}{3\,\pi} \cong 2.4.$$
$$(4.9)$$

A characteristic feature of Eqns (4.8) and (4.9) is the presence of the "cutoff frequency". In the case of the linear electron density distribution, this frequency depends upon the reduced atomic radius x_0. The "cutoff frequency" appears as a result of the radial electron density being bounded within some radius near the nucleus in the linear and exponential screening models, so there is the radiation frequency at which the plasma resonance conditions (2.1), (4.6) are not satisfied.

In Figure 3 we show the results of calculations of the photoeffect cross sections as a function of the reduced frequency obtained in these models. As is seen from this figure, the Thomas – Fermi and Lenz – Jensen distributions yield similar results for the photoeffect cross sections. A slight difference is that in the low frequency region the Lenz – Jensen model gives a somewhat lower value for the cross section than the Thomas – Fermi model. This is due to the already mentioned realistic decrease of the Lenz – Jensen electron density at large distances from the nucleus. The photoionization cross section obtained in the exponential screening model approximation (4.8) has a sharp maximum at $\omega_{\max}^{(\exp)} \cong 8.8\, Z$ eV.

More realistic electron distributions that take into account the atomic shell-like structure, such as, for example, the Hartree – Fock electron densities, leads to the appearance of the characteristic "oscillations" in the spectral photoeffect cross section around the "mean line" determined by the Thomas – Fermi and Lenz – Jensen distributions.

In paper [29], the atomic photoeffect cross sections were calculated for different modifications of the Brandt – Lundqvist approximations like Eqn. (2.8) and various forms of the dielectric permeability, including the local dispersion taken in the form:

$$\varepsilon(q,\, \omega,\, r) = \frac{\omega_p^2(r)}{\omega^2 - \frac{3}{5}\, v_{\mathrm{F}}^2(r)\, q^2}\, . \tag{4.10}$$

Here $v_{\mathrm{F}}(r) = (3\,\pi^2\, n(r))^{1/3}$ is the local Fermi velocity of atomic electrons. This account of the local dispersion of the atomic dielectric permeability somewhat decreases the photoeffect cross section in the low frequency range and increases it at high frequencies.

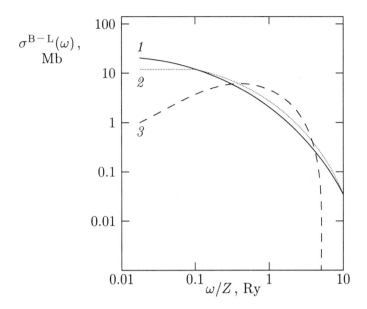

Fig. 3. The photoeffect cross section in the Brandt – Lundqvist local plasma approximation for different statistical models of the electron density: *1* — the Thomas – Fermi model, *2* — the Lenz – Jensen model, *3* — the exponential screening model.

Besides, in [29] photoabsorption cross sections for ions with different ionization degree were also computed using the Lenz – Jensen distribution. Analysis carried out in [29] revealed a significant decrease of the atomic photoeffect cross section at frequencies $\omega < Z$ Ry with increasing ionization degree (parameter $q = (Z - N_e)/Z)$), whereas at high frequencies $\omega > 3Z$ Ry the photoionization cross section is virtually independent of q.

4.2. Approximate quantum methods for photoabsorption cross section calculations

Along with essentially classical methods of accounting for inter-particle correlations in the photoabsorption considered in the previous section, some authors have developed quantum methods that take into account polarization many-particle effects. Within the frames of these methods, the photoionization cross sections were calculated using somewhat simplified approaches compared to consistent quantum-mechanical treatment, such as RPEA [15].

One such method, based on the local electron density functional formalism, was used in paper [27] for numerical calculations of photoabsorption in the noble gas atoms and their static polarizabilities. The calculations were simplified by introducing the local effective potential to find single-particle wave functions of the system's ground state. To this aim, generally non-local exchange-correlation energy was calculated in the local density approximation according to the equalities:

$$
V_{xc}(r) = -\frac{0.611}{r_s(r)} - 0.0333 \ln\left(1 + \frac{11.4}{r_s(r)}\right), \quad \frac{4}{3}\pi r_s^3(r) = n^{-1}(r).
$$

(4.11)

As a result, the solution of the corresponding equations turned out to be not more complicated than the Hartree equation solution. Polarization-correlation effects were taken into account by introducing the self-consistent field in the form of the sum of external and induced fields, which was a solution to the corresponding integral equation. Note that in the RPEA calculations [15], the effective dipole moment $\mathbf{D}(\omega)$, describing many-particle correlations, was also a solution to an integral equation.

The results of photoabsorption cross section calculations made in [27] proved in excellent agreement with the existing experimental data. In addition, they demonstrated an important role of polarization many-particle effects in photoionization of atoms with completed electronic shells. These effects lead (with exception of the neon atom) to a significant shift of the photoionization cross section maximum toward high frequencies, compared to the independent electron approximation, in which the location of the maximum virtually coincides with the threshold photon energy. For example, the photoionization cross section maximum for the xenon atom near the $4d$ threshold is shifted by about 2.5 Ry toward high frequencies. In this case, there is no strong resonance due to (in the framework of the single-particle treatment) the transition from the $4d$-subshell to virtual f-state in the continuum.

The shift of the maximum photoionization cross section mentioned above is the consequence of a redistribution of the oscillator strengths of atomic transitions from a near-threshold region to the high frequency part of the spectrum due to the atomic core polarization. This polarization leads to a specific screening/de-screening of the external field. An analysis of frequency – space dependencies of the local self-consistent field indicated that there is a strong screening of the external field in the low frequency wing of a photoabsorption line at small distances from the nucleus, so that the vector of the self-consistent field strength is directed oppositely to that of the induced field. It is interesting that for all frequencies of the ionizing radiation the "switching" of the screening regime to the "de-screening" as the distance from the nucleus increases occurs at the local electron density maximum of the atomic subshell that mostly contributes to the process cross section.

It is interesting to notice that the local electron density method used in paper [27] predicts a smaller (by several electron – Volts) value of the photoeffect threshold compared to what is actually observed. It is important to stress that here the sum rule for the photoabsorption cross section is satisfied as the "non-physical" contribution to the cross section from the continuum is compensated by ignoring the contribution from the discrete spectrum adjacent to the photoionization threshold. This fact is much more relevant to the above-mentioned variants of the classical description of the atomic photoeffect. As seen from Fig. 3, the Thomas – Fermi and Lenz – Jensen models for the atomic electron density give photoionization cross section largely extending to the low frequency region, although the sum rule for the corresponding cross sections is respected. Within the frames of these statistical models, the discrete atomic spectrum is totally absent so the "non-physical" region of the continuum below the photoionization threshold models, to some degree, the contribution from unaccounted-for bound states.

To conclude this section, we consider a simple quantum-mechanical model for atomic photoeffect, which admits an analytical representation for the process [49]. From the formal viewpoint, this approach is based on the approximate operator equality

$$e^{-i(H_0+\Delta_1)t}\, e^{i\,H_0\,t} \approx e^{-i\,\Delta_1\,t}, \quad \Delta_1 = \frac{1}{r^2}, \tag{4.12}$$

which yields the following cross section:

$$\sigma_{\mathrm{ph}}(\omega) \approx \frac{2\,\pi\,Z^2}{3\,c\,\omega} \int\limits_{-\infty}^{+\infty} dt\, \langle\psi|\, e^{-i\,\Delta_1\,t}\,|\psi\rangle\, e^{i\,\omega\,t}. \tag{4.13}$$

The representation of the cross section in form (4.13) is named in [49] the "hybrid" approximation. It is quantum mechanical because of the general operator approach and at the same time has some classic features since the approximate commutation of the operator's exponents (4.12) is used.

Note that Eqn. (4.13) can be rewritten via the electron density utilizing the following substitution:

$$|\psi(r)|^2 \to 4\,\pi\,r^2\,n(r). \tag{4.14}$$

After integrating over time the remaining integral can be taken due to the presence of the delta-function with the net result in the form:

$$\sigma_{\mathrm{ph}}(\omega) = \frac{8\,\pi^3\,Z^2}{3\,c}\, \frac{1}{\omega^{7/2}}\, n\left(r = \frac{1}{\sqrt{\omega}}\right). \tag{4.15}$$

Equation (4.15), in particular, yields the hydrogen-like high frequency approximation for the photoionization cross section if $n(r \to 0) \to$ const.

Thus, like in the Brandt–Lundqvist approximation (4.4), the photoionization cross section in the Rost hybrid approximation [49] is an electron density functional. However, in this case the characteristic radiation scale r_ω is determined not by plasma resonance condition (2.1) but by the difference of the atomic Hamiltonians with orbital numbers differed (in accordance with the dipole selection rules) by unity:

$$\omega = H_1(r) - H_0(r). \tag{4.16}$$

Equation (4.16) directly follows from Eqn. (4.12) and the energy conservation law.

Based on Eqn. (4.16), we can suggest a physical interpretation to the Rost approximation. This equation implies that the photon absorption occurs with a fixed electron's coordinate, like in the Born – Oppenheimer approximation where the coordinates of molecular nuclei do not alter by electron transitions. Note that Eqn. (4.12) is the mathematical expression of this fact. Thus the hybrid Rost approximation can be considered as the generalization of the adiabatic principle on the case of electronic transitions in atoms.

Figure 4 shows the ratio of the photoionization cross section for the ground state of the hydrogen atom calculated using the Rost formula (curve 1) and in the Kramers approximation (curve 2), to the Sommerfeld cross section. This figure implies that the quasi-classical Kramers approximation describes the photoeffect cross section near the threshold somewhat better than the Rost model, and vice versa at high frequencies. Indeed, in the high frequency limit, the Kramers approximation provides inaccurate asymptotic (ω^{-3} instead of $\omega^{-7/2}$)), whereas the ratio of the Rost result to the exact value in this spectral range is $\pi/2\sqrt{2} \approx 1.11$.

It is important to emphasize that in contrast to the Brandt – Lundqvist approximation, the Rost model does not ensue the sum rule for the photoabsorption cross section. For example, in the case of the hydrogen atom, the corresponding integral over the frequency yields two times as large value. Note that the maximum of the hydrogen atom photoabsorption in the Rost approximation lies at the frequency $\pi/2\sqrt{2} \approx 1.11$ a.u., which is substantially less than the first excitation potential, and the maximum cross section amounts to $\sigma_{ph,max}^{(R)} \cong 1.127$ a.u.

Photoabsorption formula (4.13) was generalized on the case of helium atom in paper [49]. A comparison of the obtained result with experimental data showed that the relative error for moderate photon energies is within the 5% limits.

Polarization effects are significant in the photoionization of negative ions due to a large polarizability of the ionic core, i.e. of the neutral atom. This issue is considered in review [50]. Here we briefly consider a quantum-mechanical calculation of this phenomenon for a negative ion of the alkaline metal, when the polarizability of the atom-core is especially large. The contribution of polarization effects into this process can be accounted for using the following effective potential [51]:

$$V_{\text{eff}}(\mathbf{r},\, t) = \frac{\alpha(\omega)\, \mathbf{E}\, \mathbf{r}}{r^3}\left[1 - \exp\left(-\left(\frac{r}{r_0}\right)^3\right)\right]\cos(\omega\, t), \qquad (4.17)$$

where $\alpha(\omega)$ is atom's dynamic polarizability, r_0 is the size of its outer orbit. The factor in the square brackets in Eqn. (4.17) describes the polarization interaction decrease at small distances. To calculate photoionization cross section using Eqn. (4.17) and the conventional "direct" interaction potential of the electromagnetic field with the target's electron, it is convenient to make use of an analytical expression for the wave function of the negative ion's external electron obtained by approximating variational results [52].

The spectral photoionization cross section of the negative lithium ion ($I = 6.18$ eV, $\alpha(0) = 162$ a.u. [53]), calculated in the plane-wave approximation for the ionized electron, is shown in Fig. 5 with and without taking into account polarization effects. It is clear that

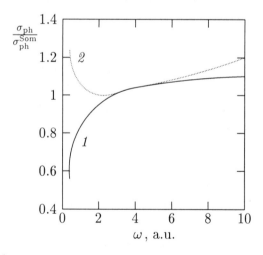

Fig. 4. The ratio of the photoionization cross section from the hydrogen atom ground state in various approaches: curve *1* for the Rost approximation [49], curve *2* for the Kramers approximation, to the Sommerfeld cross section.

the latter effects strongly decrease the cross section at frequencies of the order of the maximum frequency, which is linked to the opposite direction of the field induced by the ionic core with respect to the external field direction.

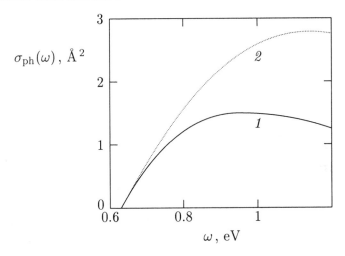

Fig. 5. The photoionization cross section for the negative ion of lithium calculated with account for polarization of the core (curve *1*) and ignoring polarization effects (curve *2*).

5. Bremsstrahlung and photorecombination of moderate energy electrons on multielectron atoms and ions

Bremsstrahlung and photorecombination of electrons during their scattering on atoms and ions are fundamental processes of plasma physics and play an important role in other fields of physics and numerous technical applications. As a rule, these phenomena have been considered without taking into account the polarization channel. However, in some cases its contribution can be quite essential. First of all, this relates to processes involving multielectron targets when a consistent quantum-mechanical calculation is very complicated due to the many-particle character of the problem. However, it is in this case that the application of approximate methods based on plasma models for atom (ion) polarizability and on the statistical method for target's electron core description becomes very effective. Moreover, bremsstrahlung of moderate energy electrons, which are natural for plasma, can be successfully treated by well-developed methods of the so-called Kramers electrodynamics. Such methods used for the static channel calculations have demonstrated a good accuracy combined with physical clarity. The generalization of these methods to the polarization channel enables one to account for effects of the IP penetration into the target's core and to find a universal (for all nuclear charges) description for the static and polarization radiative mechanisms based on one ground.

5.1. Radiative losses of electrons in scattering on neutral atoms

Radiative losses of electrons in a continuum spectrum in scattering on heavy atoms are due to their bremsstrahlung in atom's potential field (below in this Section we shall ignore the losses due to excitations of atomic electrons and corresponding to the discrete radiation spectrum). A characteristic feature of these processes is the incident electron penetration into the atomic core which increases the effective charge interacting with the electron. This effect leads to the appearance of the characteristic increase of the bremsstrahlung spectrum frequency dependence, unlike the decreasing spectral dependence in the pure Coulomb field, see review [47]. This phenomenon is important in making diagnostic and estimating radiative losses in plasma with heavy ions which have a considerable electronic core. In fact, here we deal with the interaction between electrons with energy from 0.5 to 10 keV and atoms with nuclear charge > 20.

Calculations of the bremsstrahlung (BR) in a static atom's potential have been performed by various methods: within the frames of the Born approximation [54], using the semi-classical method [45], by numerically solving the Schrödinger equation in the Thomas–Fermi potential [55, 56], by the method of the self-consistent field [57], or by the quasi-classical method [46]. In the last case it was demonstrated that for moderate energy electrons (common for plasma) the classical approximation for the atom potential is precise. Moreover, in calculating spectra at high frequencies which are responsible for the largest electron energy losses, it is possible to use the so-called "rotation approximation" (RA) which corresponds to taking into account radiation from the most distorted part of the electron's trajectory in the atomic potential. In this approximation, the bremsstrahlung spectrum is entirely determined by the dynamics of electron scattering and is expressed as a functional of that potential, see [46, 47] for more detail. A thorough comparison of classical and quantum calculations has confirmed the high accuracy of the classical method used, which is about 5% for a pure Coulomb potential and varies within the 20% range for complex ions depending on the ionic core structure.

In spite of generally good correspondence between the specified theories based on the static potential of the electron – atomic interaction, a comparison of the results of calculations [57] with experimental data [58] reveals a systematic discrepancy for atoms with large nuclear charges ($Z > 60$), mostly notable at low frequencies.

This discrepancy with experiment can exist for a number of reasons including the contribution due to the polarization channel related with the dynamic polarization of the atomic core, which has been ignored for the ordinary (static) bremsstrahlung. Numerous calculations of the polarization bremsstrahlung (PBR) carried out in recent years [9] indicate that the contribution due to this process can be comparable to (and even in excess of) the static channel contribution. In paper [30], a detailed quantum-mechanical calculation of PBR on a multielectron atom in the wide spectral range has been performed for krypton using the method of random phases with exchange. However, until the present time, there are no extensive calculations of the PBR spectra similar to those carried out in [46] for the static potential. Such calculations are possible using an atom plasma model and the quasi-classical approximation like that employed in [46] for static bremsstrahlung (SBR). Thus the direct comparison between PBR and SBR becomes possible for all kinds of atoms in a wide range of emitted quanta frequencies and the incident particle (IP) energies.

The expression for the spectral effective bremsstrahlung in the static channel was first derived in [46] with the form (in all cases we use the atomic units):

$$\left(\frac{d\kappa}{d\omega}\right)^{\text{rot}}_{\text{st}} = \frac{8\,\pi\,A}{3\,c^3\sqrt{2\,E}} \int_0^\infty (f_{\text{st}}(r))^2 \sqrt{1 + \frac{|U(r)|}{E}} \, \delta\left(\omega - \omega_{\text{rot}}(r)\right) r^2 \, dr \,.$$

(5.1)

Here A is the normalization factor, $U(r)$ is the atomic potential,

$$f_{\text{st}}(r) = -dU/dr = Z_{\text{ef}}(r)\Big/r^2$$

(5.2)

is the ordinary "static" force acting on the incident electron, $Z_{\text{ef}}(r)$ is the effective charge of the atom at a distance r from the nucleus, E is the initial kinetic energy of the incident particle, and $\omega_{\text{rot}}(r)$ is the "rotation" frequency defined by the equality

$$\omega_{\text{rot}}(r,\, E) = \frac{\sqrt{2\,(E + |U(r)|)}}{r} \,.$$

(5.3)

The rotation frequency $\omega_{\text{rot}}(r, E)$ arises in the quasi-classical limit for matrix elements that determine the PBR spectrum in the atomic potential. The inclusion of the delta-function in Eqn. (5.1) means the prevalence of matrix elements with the frequency difference $\omega - \omega_{\text{rot}}$ over the corresponding matrix elements containing their sum, see [46]. Since the same quasi-classical wave functions of the incident electron appear in the PBR calculations, it is natural to use the rotation approximation for the polarization channel too.

The simplest variant of the rotation approximation in the PBR theory consists in substituting the "static" force in Eqn. (5.1) by the "polarization" force [11]. The polarization force that depends on frequency reads

$$f_{\text{pol}}(r,\, \omega) = \frac{N_{\text{pol}}(r,\, \omega)}{r^2},$$

(5.4)

where $N_{\text{pol}}(r,\, \omega)$ is the effective charge of the atomic core, this charge responsible for radiation at frequency ω via the polarization channel. In the local electron density model [25] can be expressed as

$$N_{\text{pol}}(r,\, \omega) = \omega^2 \left| \int_0^r \beta(r',\, \omega)\, 4\pi\, r'^2 \, dr' \right|.$$

(5.5)

Here $\beta(r, \omega)$ is the spatial density of target's dynamic polarizability for which we use below the Brandt – Lundqvist approximation [3] corresponding to a static plasma model of atom:

$$\beta(r, \omega) = \frac{n(r)}{4\pi\, n(r) - \omega^2 - i0}.\tag{5.6}$$

Here $n(r)$ is the local electron density of the atomic core. The existence of the current radius (the distance from the nucleus) in the upper limit of integration in Eqn. (5.5) describes the incident electron penetration into the target's core and related effects.

It should be noted that the inclusion of polarization force (5.4) – (5.6) is rather conventional because it is defined both by the real and imaginary parts of polarizability (5.6). In this connection we should point that the possible interference of the static and polarization radiative channels can be connected with the real part only. Simple considerations indicate a comparative smallness of the interference effects, which is confirmed by calculations [30]. As follows from Eqn. (5.5), this is due to polarization effects being proportional to the square of the emitted frequency ω^2. So the contribution of the real part of the polarizability can be appreciable for sufficiently high ω. However, owing to the incident electron penetrating into the core, core effective charge (5.5) decreases whereas the effective charge Z_{ef} increases, so the interference terms are numerically small. Thus, the smallness of the interference contribution is due to effects of the incident electron penetration into target's core. These effects are quite essential inside the considered interval of frequencies and energies. We should note that the interference of the static and polarization channels could be important when the penetration can be neglected [59] or is not too essential [60].

Method (5.5) – (5.6) can be termed the local plasma frequency approximation. It is physically applicable for treatment of polarization effects within the frames of statistic models of atom.

Substituting Eqn. (5.4) into (5.1), we obtain the following expression for the radiative spectral energy losses through the polarization channel in RA

$$\left(\frac{d\kappa}{d\omega}\right)^{\text{rot}}_{\text{pol}} = \frac{8\pi A}{3\,c^3\sqrt{2E}} \int\limits_0^\infty (N_{\text{pol}}(r,\omega))^2$$

$$\times \sqrt{1 + \frac{|U(r)|}{E}}\, \delta(\omega - \omega_{\text{rot}}(r))\, r^{-2}\, dr. \qquad (5.7)$$

Here it is suitable to note that the electrostatic interaction of electrons with each other is small in comparison with their interaction with the nucleus, which exceeds the former by seven times for the Thomas – Fermi model (see the problem to Sect. 70 in [2]). This fact allows us to use in the PBR consideration the same trajectory of the incident electron as in the SBR.

The presence of the delta-function in Eqns (5.1) and (5.7) allows direct integration. Then with the help of Eqn. (5.3) we obtain for the statistic channel (cf. [46]):

$$\left(\frac{d\kappa}{d\omega}\right)^{\text{rot}}_{\text{st}} = \frac{8\sqrt{3}\pi\omega}{3\,c^3\sqrt{2E}}\, \frac{(Z_{\text{ef}}(r_{\text{ef}}(\omega,E)))^2\sqrt{1 + \frac{|U(r_{\text{ef}}(\omega,E))|}{E}}\, r^2_{\text{ef}}(\omega,E)}{\omega^2 r^3_{\text{ef}}(\omega,E) + Z_{\text{ef}}(r_{\text{ef}}(\omega,E))}. \qquad (5.8)$$

Analogically, for the polarization channel we have:

$$\left(\frac{d\kappa}{d\omega}\right)^{\text{rot}}_{\text{pol}} = \frac{8\sqrt{3}\pi\omega}{3\,c^3\sqrt{2E}}\, \frac{(N_{\text{pol}}(r_{\text{ef}}(\omega,E),\omega))^2\sqrt{1 + \frac{|U(r_{\text{ef}}(\omega,E))|}{E}}\, r^2_{\text{ef}}(\omega,E)}{\omega^2 r^3_{\text{ef}}(\omega,E) + Z_{\text{ef}}(r_{\text{ef}}(\omega,E))}. \qquad (5.9)$$

In equations (5.8) and (5.9) we have introduced the characteristic emission radius in the rotation approximation $r_{\text{ef}}(\omega, E)$ [46, 47] as defined from the equation:

$$2\,(E + |U(r)|) = \omega^2\, r^2. \qquad (5.10)$$

The physical sense of Eqn. (5.10) is that the emission in the approach under study is largely determined by the distance from the target's core at which the emission frequency coincides with the angular rotation velocity of a classical electron in the atomic field at the point of maximum approach.

The expression for the spectral R-factor in the rotation approximation follows from Eqns (5.8) and (5.9):

$$R^{(\text{rot})}(\omega,\, E) = \left\{ \frac{d\kappa_{\text{pol}}(\omega,\, E)}{d\kappa_{\text{st}}(\omega,\, E)} \right\}^{(\text{rot})} = \left[\frac{N_{\text{pol}}(r,\, \omega)}{Z_{\text{ef}}(r)} \right]^2_{r=r_{\text{ef}}(\omega,\, E)}.$$

$$(5.11)$$

The total energy losses due to bremsstrahlung in the static and polarization channels can be obtained by integrating Eqns (5.8) and (5.9) over frequency up to the initial kinetic energy E. On the other hand, this integration can be performed in Eqns (5.1) and (5.7) containing the delta-function. Then we have for the total effective emission in each channel:

$$\kappa_{\text{st}} = \frac{8\sqrt{3}\,\pi}{3\,c^3\sqrt{2E}} \int\limits_{r_{\text{ef}}(E,E)}^{\infty} (Z_{\text{ef}}(r))^2 \sqrt{1 + \frac{|U(r)|}{E}}\, r^{-2}\, dr, \qquad (5.12)$$

$$\kappa_{\text{pol}}^{\text{rot}} = \frac{8\sqrt{3}\,\pi}{3\,c^3\sqrt{2E}} \int\limits_{r_{\text{ef}}(E,E)}^{\infty} (N_{\text{pol}}(r,\, \omega_{\text{rot}}(r,\, E)))^2 \sqrt{1 + \frac{|U(r)|}{E}}\, r^{-2}\, dr.$$

$$(5.13)$$

Thus, if the target potential and electron density of its core are known, Eqns (5.8), (5.9), (5.12), and (5.13) (with account for Eqns (5.3), (5.5), and (5.6)) give the solution to the problem for quasi-classical electrons in the general form. This method will be used below for calculation of the spectral and total energy losses within the frames of the static model of atom.

As is evident from Eqns (5.9) and (5.13), the key quantity (which defines the emission in polarization channel in the given approximation) is the frequency-depending effective polarization charge $N_{\text{pol}}(r, \omega)$; we shall examine its properties below.

First we note the general relationships for the polarization charge in the local electron density approximation that entails from Eqns (5.5) and (5.6). It is quite clear that N_{pol} has the correct high frequency asymptotic (that also follows from the quantum-mechanical expression for polarizability):

$$N_{\text{pol}}^{\infty}(r) = N_{\text{pol}}(r,\, \omega \to \infty) = N_{\text{e}}(r) = \int\limits_{0}^{r} 4\,\pi\, n(r)\, r^2\, dr. \qquad (5.14)$$

$N_e(r)$ is the number of atomic electrons in a sphere of radius r. It is natural that for $r > R_0$ (R_0 is the atomic size) the value of $N_e(r)$ is equal to the total number of electrons in the target core N_{0e}.

In the opposite low frequency limit we have from Eqns (5.5) and (5.6):

$$N_{\text{pol}}^{(0)}(r, \omega) = \omega^2 \left[\theta(r - R_0) \frac{R_0^3}{3} + \theta(R_0 - r) \frac{r^3}{3} \right]. \qquad (5.15)$$

We note that expression for the target's static dipole polarizability, which follows from Eqn. (5.15), of target $\alpha_0 = \omega^{-2} N_{\text{pol}}^{(0)}(r > R_0, \omega) = R_0^3/3$ gives a reasonable accuracy for atoms and ions with closed shells and sufficiently large number of bound electrons $N_{0e} \geq 30$ if the size of atom (ion) is calculated in the Thomas – Fermi – Dirac model [25].

In the statistic models, the electron density of a neutral atom's core that defines the polarization charge (see Eqns (5.5), and (5.6)), can be presented in the form [2]:

$$n(r) = Z^2 f(r/r_{\text{TF}}). \qquad (5.16)$$

Here $r_{\text{TF}} = b/Z^{1/3}$ is the Thomas – Fermi radius, Z is the atomic "charge", $b = \left(\frac{9\pi^2}{128} \right)^{1/3}$. The form of the function $f(x)$ is dictated by the statistic model choice. For example, in the Thomas – Fermi model we have [2]:

$$f_{\text{TF}}(x) = \frac{1}{4\pi b^3} \left(\frac{\chi(x)}{x} \right)^{3/2} \qquad (5.17)$$

where $\chi(x)$ is the Thomas – Fermi function. The expression for $f(x)$ in the Lenz – Jensen statistic model which better describes the behavior of the electron density at large distances from the nucleus reads [1]:

$$f_{\text{LJ}}(x) \cong 3.7 \exp\left(-\sqrt{9.7x} \right) \frac{(1 + 0.26\sqrt{9.7x})^3}{9.7x^{3/2}}. \qquad (5.18)$$

We note that for $x \leq 1$, Eqns (5.17) and (5.18) give almost coincident results.

Substituting Eqns (5.16) and (5.17) into Eqns (5.5) and (5.6) yields the following expression for the polarization charge in the Thomas – Fermi model:

$$N_{\text{pol}}(r, \omega, Z) = Z g \left(\frac{r}{r_{\text{TF}}}, \frac{\omega}{Z} \right) . \tag{5.19}$$

Here we have introduced the universal function $g(x, \nu)$:

$$g(x, \nu) = \nu^2 \left| \int_0^x \frac{\chi^{3/2}(x')\sqrt{x'}dx'}{b^{-3}(\chi(x')/x')^{3/2} - \nu^2 - i0} \right| . \tag{5.20}$$

which is the polarization charge normalized to the total number of atomic electrons as a function of the dimensionless distance $x = r/r_{\text{TF}}$ and the reduced frequency $\nu = \omega/Z$. Equations (5.19) and (5.20) provide a universal representation for the polarization charge in the Thomas – Fermi model.

Note that Eqns (5.19) and (5.20) generalize a one-parametric similarity law for the dipole polarizability of the Thomas – Fermi atom first obtained in [46] for the case of two variables. The dipole limit for N_{pol} is obtained from Eqns (5.19) and (5.20) if in Eqn. (5.20) the upper limit of integration is set to infinity: $g^{\text{dip}}(\nu) = g(x \to \infty, \nu)$.

Comparison of calculations made in the frames of the considered method for the real and imaginary parts of the dipole polarizability of the krypton atom (multiplied by the frequency square) with quantum-mechanical calculations carried out in the approximation of random phases with exchange taken from [30], indicates that in the case of the Thomas – Fermi atom, the local plasma frequency method describes on average the exact spectral dependence that accounts for the shell structure of atom. For Slather's electron density, the method in use leads to the appearance of maxima and minima due to ionization of electronic subshells. Here, however, the universality of description, which is typical for the Thomas – Fermi atom, is lost. So, we can conclude that the approximation in use gives a reasonable accuracy for the atomic dipole polarizability value but does not describe, naturally, the quantum-mechanical properties of the atomic electron distribution.

In what follow we shall examine the case for the polarization charge calculated in RA, $N_{\text{pol}}^{(rot)}(\omega, E)$, because it accounts for the

effects of the IP penetration into the target's core which are essential
for the quasi-classical electrons. The corresponding expression can
be obtained using Eqn. (5.5) and the equation:

$$N_{\text{pol}}^{(rot)}(\omega, E) = N_{\text{pol}}(r_{\text{ef}}(\omega, E), \omega) \qquad (5.21)$$

Here $r_{\text{ef}}(\omega, E)$ is the solution to equation (5.10). Hence in the rotation approximation we obtain for the normalized polarization charge:

$$g^{(rot)}(\nu, \varepsilon) = g(x_{\text{ef}}(\nu, \varepsilon), \varepsilon) \qquad (5.22)$$

Here $\varepsilon = bE/Z^{4/3}$ is the reduced energy; $x_{\text{ef}}(\nu, \varepsilon)$ is the solution to
Eqn. (5.10) rewritten in terms of parameters ν and ε.

Spectral energy losses of quasi-classical electrons due to the ordinary (static) channel in scattering on a Thomas – Fermi atom were
calculated in [46] both with the use of RA (inside its applicability
region $\nu > 3\varepsilon$), and in the low-frequency interval ($\nu < 3\varepsilon$)) using
linear interpolation to the "transport" limit. In paper [46], the ob-
tained results were compared with the results of quantum-mechanical
calculation [57] which demonstrated a high accuracy of the approximations used. Within the framework of RA, we shall obtain here
a universal expression for the spectral bremsstrahlung losses due to
the polarization channel in quasi-classical electron scattering on a
Thomas – Fermi atom. This can be conveniently done in terms of
R-factor (5.11), which is the ratio of contributions due to polarization and static radiative mechanisms. By passing in Eqns (5.4)
and (5.11) to the reduced variables, it is easy to obtain the following
expression:

$$\widetilde{R}_{\text{TF}}^{(rot)}(\nu, \varepsilon) = \left[\frac{g(x, \nu)}{\chi(x) + x|\chi'(x)|} \right]^2_{x = x_{\text{ef}}(\nu, \varepsilon)}, \quad 3t < \nu < \nu_{\text{hf}} \leq 10\,.$$

$$(5.23)$$

Here the prime stands for the differentiation with respect to the argument and $g(x, \nu)$ is the normalized polarization charge defined by
Eqn. (5.20). The maximum reduced frequency ν_{hf} follows from the
energy conservation law and is a function of the reduced initial en-
ergy and the nuclear charge: $\nu_{\text{hf}} = \left(\sqrt[3]{Z}/b \right) \varepsilon$. The transition to the

ordinary frequency and energy is performed with the help of equation:

$$R_{\text{TF}}^{(\text{rot})}(\omega, E) = \widetilde{R}_{\text{TF}}^{(\text{rot})}\left(\frac{\omega}{Z}, \frac{Eb}{Z^{4/3}}\right), \qquad \frac{b}{Z^{1/3}}E < \omega < E. \quad (5.24)$$

Thus, Eqns (5.23), (5.24), and (5.20) and the definition $x_{\text{ef}}(\nu, \varepsilon)$ give a universal (correct for all nuclear charges) representation for spectral R-factor in the approach considered. These relations should be completed with the expression for spectral losses via the static channel (5.8); for the Thomas – Fermi atom they expressed through reduced variables:

$$\left(\frac{d\kappa}{d\omega}\right)_{\text{st}}^{\text{rot}} = \frac{8\sqrt{3}\pi Z^{2/3}b^{5/2}\nu}{3c^3\sqrt{2\varepsilon}} \left.\frac{x^4[(\chi/x)']^2\sqrt{1+\chi/(x\varepsilon)}}{b^3 x\nu^2 + |(\chi/x)'|}\right|_{x=x_{\text{ef}}(\nu,\varepsilon)}. \quad (5.25)$$

An equivalent to Eqn. (5.25) expression was obtained for the first time in [46] in terms of the Gaunt factor for RA. In the Coulomb ($\chi(x) = 1$) and Kramers ($x_{\text{ef}}^{-1} > \varepsilon$) limits, one can obtain from Eqn. (5.24):

$$\left(\frac{d\kappa}{d\omega}\right)_{\text{st}}^{\text{rot,Coul}} = \left(\frac{d\kappa}{d\omega}\right)^{\text{Kramers}} = \frac{8\pi}{3\sqrt{3}c^3}\frac{Z^2}{E}. \quad (5.26)$$

Spectral emission in polarization channel is obtained from Eqn. (5.25) by the substitution $x^2(\chi/x)' \to g(x,\nu)$ which follows from Eqn. (5.23):

$$\left(\frac{d\kappa}{d\omega}\right)_{\text{pol}}^{\text{rot}} = \frac{8\sqrt{3}\pi Z^{2/3}b^{5/2}\nu}{3c^3\sqrt{2\varepsilon}} \left.\frac{g(x,\nu)^2\sqrt{1+\chi/(x\varepsilon)}}{b^3 x\nu^2 + |(\chi/x)'|}\right|_{x=x_{\text{ef}}(\nu,\varepsilon)}. \quad (5.27)$$

Here we also present the corresponding expression, obtained in [61] in the Born – Bethe approximation, in terms of the reduced variables:

$$\left(\frac{d\kappa}{d\omega}\right)_{\text{pol}}^{\text{B-B}} = \frac{8}{3c^3}\frac{Z^{2/3}}{\varepsilon}|g_{\text{dip}}(\nu)|^2 \ln\left(\frac{1}{\nu}\sqrt{\frac{2\varepsilon}{b^3}}\right), \qquad \varepsilon > \frac{1}{8bZ^{2/3}}. \quad (5.28)$$

Here $g_{\text{dip}}(\nu) = g(x \to \infty, \nu)$ is normalized polarization charge (5.20) in dipole approximation. Equation (5.28) implies that, contrary to RA, in the Born – Bethe approximation there is an upper frequency limit: $\nu < 1.7\sqrt{\varepsilon}$.

Figure 6 presents the dependences of the electron effective emission on the Thomas – Fermi atom on the reduced frequency in polarization and static channels for two values of the IP initial energy calculated within the frames of various approximations. Curves *1* and *3* illustrate the polarization channel calculated in the rotation approximation and the Born – Bethe approximation (5.28) respectively. In calculating dependence *2* (corresponding to the static channel), in the low frequency interval the Gaunt factor $g_0(\varepsilon)$ for the Thomas – Fermi atom was used in the "transport" limit. The corresponding interpolation for the function $g_0(\varepsilon)$ was obtained using data from paper [62]. Figure 6 indicates that the polarization emission calculated in the rotation approximation at $\nu < \varepsilon$ virtually coincides with the results obtained in the Born – Bethe approximation. Note that the inequality $\nu < \varepsilon$ can be rewritten in the form: $\omega < bE/Z^{1/3}$. Hence for large Z it corresponds to the applicability condition for the low frequency approximation in the BR theory. In this case, the electron scattering on neutral atom is weakly perturbed by the radiative process and for the PBR calculation its motion can be considered as uniform and rectilinear even for slow electrons [59]. This justifies the applicability of the Born – Bethe approximation for polarization channel at low frequencies. Thus, owing to a good conjugation of curves *1* and *3*, the rotation approximation will be applied for calculation of polarization emission in the entire spectral interval under consideration. The decrease of $(d\kappa/d\omega)_{\text{pol}}^{\text{rot}}$ with frequency (illustrated in Fig. 6) is due to effects of the IP penetration into the target's core. However, this fall off is not so sharp as in the Born – Bethe approximation. The initial increase of $(d\kappa/d\omega)$ with frequency is well known from the PBR theory [9] and is caused by the factor ν^2 included in polarization charge definition (5.20). So the spectral dependence of the PBR intensity has a maximum with the central frequency shifting toward high frequencies as the IP initial energy increases. This fact is in agreement with the conclusion of paper [63], in which the correlation $\omega_{\text{max}} \approx 0.8E$ was obtained in the Born approximation. Figure 6 also implies that everywhere in the considered interval of parameters the contribution due to polarization channel is less than the static contribution, with the difference growing with the IP energy.

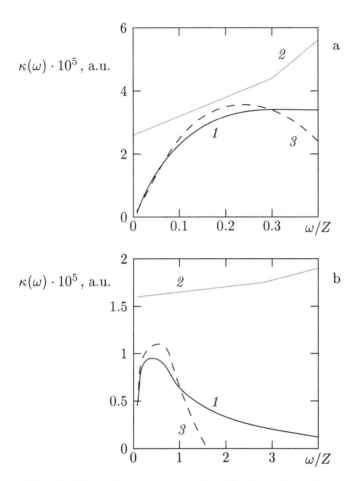

Fig. 6. The effective spectral radiation of an electron on a Thomas–Fermi atom ($Z = 60$) as a function of the reduced frequency $\nu = \omega/Z$ for different values of the incident electron reduced energy (a) $t = 0.1$; (b) $t = 1$. Curve *1* stands for the polarization channel (in the rotational approximation), curve *2* for the static channel, and curve *3* for the polarization channel (in the Born–Bethe approximation).

Figure 7 exhibits the spectral effective emission in static and polarization channels for electrons with energies 1 keV and 5 keV scattering on the krypton atom, as calculated by the present paper method (curves *1* and *3*), and using consistent quantum-mechanic methods in [30] (curves *2* and *4*). The dependencies for static channel has been obtained using Eqn. (23) and the above-mentioned procedure of linear interpolation to low frequencies $\nu < 3\varepsilon$ [46]. Note that the static BR was calculated in [30] within the frames of the distorted plane wave approximation by summing up partial contributions from different angular momenta with the use of the exact potential of the krypton atom. It is evident that both methods yield very close results for the static channel. A small difference between curves *3* and *4* at high frequencies is due to a certain freedom in the choice of the upper bound frequency for the linear interpolation. For the polarization channel our result more strongly deviates from that of paper [30] obtained using the random-phase-with-exchange approximation which includes not only individual quantum-mechanical properties of the atomic electron motion but also inter-particle correlation effects. This difference is mostly strong near the ionization potentials of electronic subshells where the real and imaginary parts of the atomic polarizability have resonance structures. For the 1-keV electrons, the broad dip in the spectral PBR intensity is due to ionization of the $3d$ subshell of the krypton atom. In the case of the 5-keV electrons, the dip on curve *2* shown in Fig. 1(b) is caused by the $2p$ subshell ionization. Its relative width is perceptibly less, so accordance between the results of the method employed and consistent quantum-mechanical calculations is better. Here we should note that in [30] (unlike the present paper), the cross-channel interference was also taken into account, which was found to have generally a minor effect on the total BR intensity. The interference effects are mostly prominent near ionization potentials of the electronic subshells and at high frequencies. Apparently, these effects are responsible for the difference between curves *1* and *2* of Fig. 2(b) for photon energies above 2200 eV. As a whole, Fig. 7 demonstrates reasonable accuracy of the method employed for the PBR calculation on a multielectron atom.

Note that the relative contribution of the polarization channel at a given reduced frequency increases with the initial electron

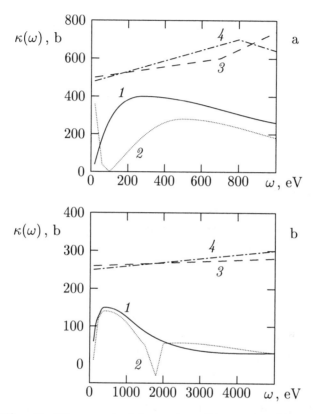

Fig. 7. The spectral intensity of bremsstrahlung radiation of an electron with energy 1 keV (a) and 5 keV (b) on the krypton atom for different radiative channels: *1* — the polarization channel (calculation of the present paper), *2* — the polarization channel with inclusion of the interference contribution (calculation in the random phase with exchange approximation [30]), *3* — the contribution from the static channel (calculation of the present paper), *4* — contribution from the static channel [30].

energy, though the maximum value of the R-factor decreases with
the IP energy. This is because within the frames of RA, the effective emission radius increases with the IP energy for a fixed emission frequency. The spectral R-factor reaches maximum at $\nu_{max} \approx$
$0.15 - 0.45$, with ν_{max} increasing with the IP energy, while for fast
electrons $\nu_{max} \approx 1$ [64]. It is interesting to note that for slow electrons ($E < I$, I is atom's ionization potential) the maximum PBR
contribution is shifted toward high frequencies [59, 63], which is because the incident particle does not penetrate into the target's core.

Equation (5.24) for the R-factor enables us to study its dependence on the nuclear charge of the Thomas – Fermi atom at fixed
values of the emission frequency and the IP energy. The corresponding calculations for various values of frequency and fixed initial energy
in the keV range indicate that the relative contribution of polarization channel increases with the nuclear charge. The R-factor proves
larger for shorter emission frequency at the same initial energy.

The growth of the relative contribution of the polarization
mechanism as the nuclear charge increases, predicted by the developed PBR theory, is in qualitative accordance with experimental results of paper [58]. The BR intensity at low frequencies was measured in [58] to exceed double the quantum-mechanical SBR calculations (ignoring polarization channel) [57] for large nuclear charges
($Z \approx 90$). At the same time, for small and moderate Z a good
accordance between experiment and the exact theory of the ordinary
(static) BR was observed. It should be noted that quantitatively this
excess is not completely explained by the PBR theory developed here
and perhaps is a consequence of contributions due to other radiative
processes (such as, for example, the two-photon BR, see paper [58]).

Figure 8 illustrates the total (static and polarization channels)
effective emission for three values of the nuclear charge of a Thomas –
Fermi atom as a function of the photon energy. It is evident that the
PBR contribution modifies the general form of the spectrum compared to purely static case (see curves *3*, *4* in Fig. 7).

Now we turn to calculation of the total radiative losses. General expression (5.12) for the total energy losses in the static channel,
first obtained in [45] within the frames of semi-classical method, can
be rewritten for the Thomas – Fermi atom in terms of the reduced

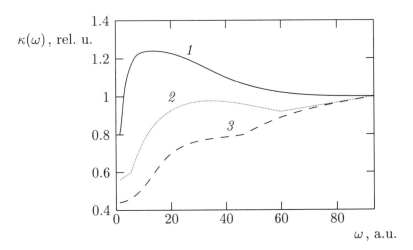

$\kappa(\omega)$, rel. u.

Fig. 8. Bremsstrahlung spectra with account for the polarization channel normalized to their values at $\omega = 90$ a.u., for an electron with energy 100 a.u. scattering on a Thomas–Fermi atom with different nuclear charges: $1 - Z = 30$, $2 - Z = 60$, $3 - Z = 90$.

variables:

$$\kappa_{st}\left(E = \frac{Z^{4/3}}{b}\varepsilon\right) = \frac{8\sqrt{3}\pi\sqrt{b}}{3c^3}\frac{Z^{5/3}}{\sqrt{2\varepsilon}}$$

$$\times \int_{x_{ef}(\nu_{hf},\varepsilon)}^{\infty}\left[\left(\frac{\chi(x)}{x}\right)'\right]^2\sqrt{1 + \frac{\chi(x)}{\varepsilon x}x^2}\,dx. \quad (5.29)$$

Here, as in Eqn. (5.23), $\nu_{hf} = \left(\sqrt[3]{Z}/b\right)\varepsilon$ is the reduced frequency corresponding to the bremsstrahlung high frequency cut-off. It is important that for expression (5.29) to be valid the RA condition $(\nu > \nu_{min}^{(rot)}(t) \cong 3\varepsilon)$ can be relaxed. Only the condition of quasi-classical motion of the scattering electron $\varepsilon \leq 1$ should be met. On the other hand the correct normalized multiplier can be obtained only by comparing the semi-classical result [45] and the quasi-classical limit of the exact quantum-mechanical expression for the spectral cross-section (see [46]).

By rewriting analogically Eqn. (5.13) for the total energy losses in the polarization channel, we obtain:

$$\kappa_{\text{pol}}\left(E = \frac{Z^{4/3}}{b}\varepsilon\right) \approx \kappa_{\text{pol}}^{\text{rot}}(t) = \frac{8\sqrt{3}\pi\sqrt{b}}{3c^3}\frac{Z^{5/3}}{\sqrt{2\varepsilon}}$$

$$\times \int_{x_{\text{ef}}(\nu_{\text{hf}},\varepsilon)}^{\infty} g(x,\nu_{\text{rot}}(x,\varepsilon))^2\sqrt{1 + \frac{\chi(x)}{\varepsilon x}}x^{-2}\,dx. \quad (5.30)$$

Here the function $g(x,\nu)$ represents normalized polarization charge (5.20); $\nu_{\text{rot}}(x,\varepsilon)$ is the reduced rotation frequency depending on the reduced distance and energy which follows from Eqn. (5.3)

$$\nu_{\text{rot}}(x,\varepsilon) = \sqrt{\frac{2}{b^3}\frac{\varepsilon + \chi(x)/x}{x^2}}. \quad (5.31)$$

Expression (5.30) for the total energy losses in polarization channel was obtained in the rotation approximation which, as was shown in the previous section, provides adequate description of PBR (contrary to SBR) at low frequencies as well.

Note that in Eqns (5.29) and (5.30) (unlike in the corresponding expressions for spectral losses (5.25) and (5.27)), in addition to the multiplier $Z^{5/3}$ in the element of integration there is an explicit dependence on the nuclear charge appears due to the lower limit of the integral over the dimensionless distance x depending on Z. Indeed, the above-mentioned lower limit is a solution to the equation:

$$t + \frac{\chi(x)}{x} = \frac{bZ^{2/3}}{2}x^2\varepsilon^2 \quad (5.32)$$

which includes Z as a parameter. Thus, the total energy losses of the quasi-classical electron scattering on the Thomas–Fermi atom exhibit no exact scaling over the reduced frequency and energy which is characteristic for the spectral energy losses. However, calculation demonstrates that the lower limit of integration in Eqn. (5.30) rather weakly depends on Z: changing the nuclear charge by two times alters the value $x_{\text{ef}}(\nu_{\text{hf}},\varepsilon)$ by about $10-15\%$. Hence, we can conclude that there is an approximate scaling in the total energy losses on the Thomas–Fermi atom.

From the viewpoint of unification of radiative processes, it is worth noting that Eqn. (5.29) can be presented in the form similar to Eqn. (5.30) if the effective (normalized) charge emitting via the static channel is introduced according to the equation $g_{st}(x) = x^2(\chi(x)/x)'$.

In the rotation approximation, the effective polarization charge increases at small distances from the nucleus by staying virtually independent of the IP energy and coincides with the radial distribution of the core electron charge. At large distances, the polarization charge becomes a decreasing function of the distance having a larger value at large IP energies. All these facts can be interpreted by taking into account the rotation frequency $\nu_{rot}(x,\varepsilon)$ dependence on the distance from nucleus and on the IP energy. At small distances, according to Eqn. (5.3), the value $\nu_{rot}(x,\varepsilon)$ is large (independent of the parameter ε), and high frequency approximation (5.3) holds for the polarization charge, in which it coincides with the radial distribution of the core electron charge. At large distances from the nucleus, the smaller the initial energy the stronger the rotation frequency decrease and the polarization charge starts diminishing, as follows from its definition (Eqn. (5.20)).

In Figure 9 we present the dependence of the Gaunt factor (the ratio between the electron effective emission on the Thomas – Fermi atom ($Z = 60$) to its Kramers analog) on the IP reduced energy for static and polarization channels of the process. Evidently, the polarization emission prevails over the static emission at low energies $\varepsilon < 0.05$, which, if expressed in ordinary units, for a given nuclear charge corresponds to $E < 360$ eV). Note that for such IP energies the Brandt – Lundqvist approximation for the target polarizability calculation become marginally applicable for the characteristic emission frequencies. Figure 9 also suggests that the total losses in polarization channel rapidly saturate with the electron energy increase in contrast to the losses in static channel that intensively increase up to the Coulomb limit. The rapid saturation of radiative polarization energy losses is explained by the electron penetration into the target core becoming more important with the IP energy and the related polarization charge decreasing. As a result, the scattering electron mostly radiates in polarization channel at relatively low frequencies

V. A. Astapenko et al

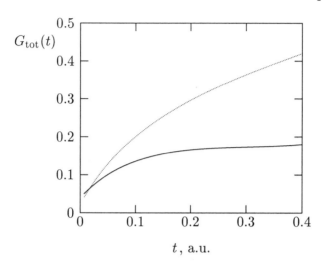

Fig. 9. The Gaunt factor for total energy losses in
scattering of a quasi-classic electron on a Thomas –
Fermi atom ($Z = 60$) for the static (the dotted
line) and polarization (the solid line) channels as a
function of the reduced energy $t = bT/Z^{4/3}$ (T is
the initial electron energy in atomic units).

so that its energy growth does not increase the total energy losses
due to the polarization radiation.

In analogy with spectral R-factor (5.23) and (5.24), we can
introduce the R-factor for total losses. In the reduced units it reads:

$$\widetilde{R}_{\text{tot}}(\varepsilon, Z) = \frac{\kappa_{\text{pol}}(\varepsilon, Z)}{\kappa_{st}(\varepsilon, Z)} \, . \tag{5.33}$$

In the dimensionful energy units we have:

$$R_{\text{tot}}(E, Z) = \widetilde{R}_{\text{tot}} \left(\frac{Eb}{Z^{4/3}}, Z \right) \, . \tag{5.34}$$

Figure 10 illustrates the dependence of R_{tot} on the initial electron
energy (in the ordinary units) for various atomic nuclear charges.
Evidently for energies $E > 1$ keV the fraction of polarization effects
in the total energy losses decreases monotonically with the IP energy.
For a fixed initial energy, the relative contribution of the polarization

channel increases with the atom – target nuclear charge, as is also the case for the spectral radiative losses.

Based of the above consideration, we can make the following conclusions. The statistic model of atom provides a high accuracy for the static BR channel. For the polarization channel, the accuracy of this model conserves on average, while near the ionization potentials of the electronic shells the divergence between the statistical model and quantum calculations [30] becomes essential. The contribution due to polarization mechanism is maximal at low frequencies $\nu \approx 0.15 - 0.45$, where the R-factor is about unity, with the central frequency of the maximum increasing with the IP energy growth. At a fixed frequency, the value of the R-factor increases with the scattering electron energy decrease at high frequencies and decreases at low frequencies. The relative value of the polarization radiation at the fixed frequency and IP energy increases with increasing atomic nuclear charge, which is in qualitative accordance with experimental data [58]. The total energy losses due to the polarization emission proves to be comparable with the losses due

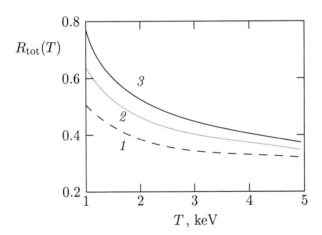

Fig. 10. The dependence of the R-factor (that characterizes the relative contribution of the polarization channel into the total energy losses) on the initial energy in quasi-classic electron scattering on a Thomas – Fermi atom of various nuclear charge: *1* — $Z = 30$, *2* — $Z = 60$, *3* — $Z = 90$.

to the ordinary (static) BR. The region of the polarization channel prevalence is defined by the inequality $E < 0.05(Z^{4/3}/b)$, where
however the characteristic emission frequencies lie at the boundary of this model applicability. In scattering of a quasi-classical
electron on a Thomas–Fermi atom, the polarization losses rapidly
saturate with the IP energy growth, owing to the scattering electron penetrating into target's core, whereas the ordinary bremsstrahlung losses go on increasing (for $\varepsilon \leq 1$). The typical value of
the polarization channel relative contribution in the total energy
losses in the 1–5 keV energy range changes from 80 to 25 per cent,
by decreasing with energy and increasing with the atomic nuclear
charge.

Thus, we can conclude that polarization effects in the bremsstrahlung of moderate-energy electrons on atom become more significant as the nuclear charge increases, as well as the frequency and IP
energy decrease.

5.2.   Core polarization effects in emission and recombination of electrons on multielectron atoms

For applied problems of high temperature plasma physics it
is necessary to calculate emission of thermal energy electrons on
ions with a core with account for the ionization state of plasma
and the polarization channel contribution. In this Section we examine the employment of plasma models for a target's electronic core,
analogous to the models applied for neutral targets, for description
of emission and recombination of plasma electrons on multielectron
ions.

The calculation of the effective emission and photorecombination of an electron on an ion with a core is based on the Thomas–
Fermi model for the density $n(r, q, Z)$ of the electron distribution in
the ion's core ($q = Z_i/Z$ is the ionization degree, Z_i is the ion
charge). The general form of the function $n(r, q, Z)$ is given by
Eqn. (5.16). The function $\chi(x, q)$ depends on the ionization degree
and can be conveniently calculated using the approximate expression

obtained by Sommerfeld and specified on in [1]:

$$\chi(x,q) = \chi_0(x) \left[1 - \left(\frac{1 + z(x)}{1 + z_0(q)} \right)^{\lambda_1/\lambda_2} \right],$$

$$z(x) = \left(\frac{x}{\sqrt[3]{144}} \right)^{\lambda_2}, \quad z_0(q) = \left(\frac{x_0(q)}{\sqrt[3]{144}} \right)^{\lambda_2}.$$

(5.35)

Here $x_0(q)$ is the reduced ion radius, $\chi_0(x)$ is the Thomas – Fermi function of the neutral atom, $\lambda_1 = (7 + \sqrt{73})/2$, $\lambda_2 = (-7 + \sqrt{73})/2$. A good approximation for the reduced ion radius can be obtained in the Thomas – Fermi – Dirac model [8]:

$$x_0(q) = 2.96 \left(\frac{1-q}{q} \right)^{2/3}, \quad 0.2 < q \le 1. \quad (5.36)$$

Approximation (5.36) is sufficient for a high-temperature plasma with electron temperature of $T > 500$ eV. For lower temperatures and correspondingly smaller ionization degrees, the parameter $x_0(q)$ can be determined from the solution of the transcendental equation $q = -x\, d\chi/dx$ [1], where $\chi(x,q)$ is given by Eqn. (5.35).

In the considered approximation the ion potential reads:

$$U(r = x r_{\text{TF}}) = \frac{Z^{4/3}}{b} \theta(x_0 - x) \left[\frac{\chi(x,q)}{x} + \frac{q}{x_0} \right] + \theta(x - x_0) \frac{q}{x} \quad (5.37)$$

and here $\theta(x)$ is the Heaviside step function.

To calculate spectral effective emission in static and polariza-tion channel within the RA frames we shall use equations from the previous section (5.8) and (5.9) with electron density of the Thomas – Fermi ion (5.16) – (5.35) and potential (5.37). It is essential that, since BR weakly depends on the IP trajectory, result (5.9) can be general-ized (as was shown in [65]) over the whole spectral interval of photons emitted. On the contrary, the interpolation of SBR (5.8) into the low frequency interval poses a problem which can be solved differently for neutral atoms and ion – targets. In the case of neutral atom linear interpolation (5.8) to the transport limit [46] proves sufficient. This procedure can not be applied for ions because the transport cross-section of an electron scattering in the Coulomb field diverges in the limit of the IP zero energies. In that case for SBR we shall use the sewing of the high and low frequency limits in the form suggested in paper [62]. The corresponding Gaunt factor for the static channel has the form:

$$
g_{\text{st}} = \frac{\sqrt{6}}{\pi} q^{\mu(\varepsilon)} \ln \left\{ \exp \left[\frac{\pi \max(q^2, g_{\text{st}}^a(\nu, \varepsilon))}{\sqrt{6} q^{\mu(\varepsilon)}} \right] \right.
$$
$$
\left. + \left[\frac{4\varepsilon^{3/2}}{1.78\sqrt{\frac{b^3}{2}} \nu q} \right]^{\frac{q^{2-\mu(\varepsilon)}}{\sqrt{2}}} \right\}. \tag{5.38}
$$

Here $\nu = \omega/Z$, $\varepsilon = Eb/Z^{4/3}$ are the reduced frequency and energy, respectively, $\mu(\varepsilon) = (1 - \ln\sqrt{\varepsilon})/2$, and $g_{\text{st}}^a(\nu, \varepsilon)$ is the Gaunt factor for SBR on the neutral Thomas–Fermi atom. The comparison of the bremsstrahlung cross-section calculated using Eqn. (5.38) with the results of consistent quantum-mechanical calculations [57] gives evidence for a good accuracy of interpolation (5.38), to within 10% as a rule.

Note that in the considered approximation the bremsstrahlung and recombination emissions smoothly transform from each other so that the high frequency spectral boundary is $\omega^{\text{h}} = E + I(Z_i, Z)$, where $I(Z_i, Z)$ is the ion–target ionization ionization.

First we consider the effective emission in both channels without temperature averaging. It is interesting to evaluate the BR contribution in the case of low energy incident electrons $E \ll I(Z_i, Z)$ in their scattering on multicharged ions when the recombination emission prevails. This situation takes place, in particular, in the experiments on the storage rings [66]. For the ion ionization potential $I(Z_i, Z)$ that determines the high-frequency emission boundary in this case, we employ the fitting obtained in [54]:

$$
I(Z_i, Z) = \frac{3(1 + Z_i)^{4/3}}{1 - 0.96 \left(\frac{1+Z_i}{Z} \right)^{0.257}}. \tag{5.39}
$$

Figure 11 demonstrates the effective SR and PR emission as computed by Eqns (5.8) and (5.9), respectively, in the frames of static

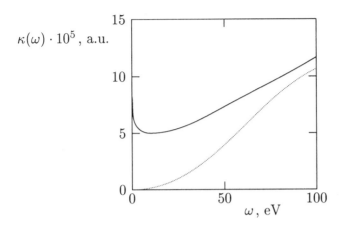

$\kappa(\omega) \cdot 10^5$, a.u.

Fig. 11. The effective radiation of a quasi-classic electron with energy 1 a.u. in scattering on the Fe^{4+} ion: the solid curve indicates the static channel, the dashed curve indicates the polarization channel.

model (5.35) for four-times ionized iron and the IP energy equal to one atomic unit. With the growth of the emission frequency the contribution of polarization channel apparently increases up to the static channel values at $\omega > 50$ eV. In this case the short-wave boundary of emission is 120 eV, in accordance with Eqn. (5.39). Since the IP initial energy is relatively small, the effect of the IP penetration into the electron core of the target does not appreciably decrease the PR intensity in the high-frequency limit, although it slightly increases the SR intensity. For $\omega < E = 1$ a.u. only bremsstrahlung is effective, while at higher frequencies the recombination emission dominates. As the IP energy decreases the short-wave boundary of emission naturally shifts toward lower frequencies. The form of the spectral dependences and the relation between them do not change principally.

The spectral dependence of the R-factor for recombination emission of the monoenergetic electrons ($E = 0.1$ a.u.) on ions with different ionization degrees is presented in Fig. 12 (for uranium as the element having the largest ion core). Naturally, the maximum value of the R-factor is obtained for the minimal ionization degree

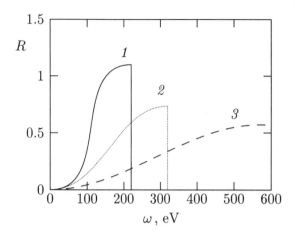

Fig. 12. The spectral R-factor for recombination radiation of electrons with energy 0.1. a.u. scattering on uranium ions ($Z = 92$) with different ionization degree: $1 - Z_i = 12$, $2 - Z_i = 15$, $3 - Z_i = 28$.

(in our case $Z_i = 12$), when polarization charge (5.5) is relatively large. In that case the contribution of the polarization channel in the recombination emission exceeds the static channel contribution at sufficiently high frequencies. As is seen from Fig. 12, for the given parameter values, there is an optimal frequency of emission at which the R-factor is maximal. For higher frequencies the relative contribution of the polarization channel in the process gets smaller due to the effects of the IP penetration into the target's electronic core (which is accompanied by the polarization charge N_{pol} decrease). The value of the optimal frequency increases with the ion charge, with the R-factor decreasing and high-frequency boundary shifting toward higher frequencies.

Now we turn to calculations of the SR and PR intensities of a high temperature plasma in the coronal equilibrium state. The temperature dependence of an ion with moderate ionization degree and given nuclear charge will be employed in the following approximate form:

$$\bar{q}(T, Z) = \frac{26}{Z} \sqrt{\frac{0.0272T[\text{a.u.}]}{1 + 0.015T[\text{a.u.}](26/Z)^2}}. \tag{5.40}$$

Equation (5.40) is a slightly altered variant of the expression given in [67]; with the 5–10% accuracy it reproduces data obtained in [68] by solving the system of equations for the coronal equilibrium. Approximation (5.40) allows the emission intensity in every channel (per one electron–ion collision) to be easily calculated, including bremsstrahlung and recombination processes with account for the temperature factor:

$$k(\omega, T, Z) = \frac{2\sqrt{2}}{\pi} T^{-3/2}$$

$$\times \int_{E_{\min}(\omega,T)}^{\infty} \frac{d\kappa}{d\omega}(\omega, E, \bar{q}(T, Z), Z) \exp\left(-\frac{E}{T}\right) E \, dE. \quad (5.41)$$

In equation (5.41), the lower limit of integration is defined by the evident equation:

$$E_{\min}(\omega, T, Z) = \max\left\{0, \omega - I(Z\bar{q}(T, Z), Z)\right\}. \quad (5.42)$$

In analogy with Eqn. (5.11), we can write for the R-factor, that describes the ratio between the temperature-averaged intensities of emission in polarization and static channels:

$$\bar{R}_T(\omega, T, Z) = \frac{k_{\mathrm{pol}}(\omega, T, Z)}{k_{\mathrm{st}}(\omega, T, Z)}. \quad (5.43)$$

Figure 13 shows the spectra of the total and bremsstrahlung emission in polarization and static channels calculated in accordance with Eqn. (5.41) for scattering of plasma electrons with a temperature of 500 eV on the tungsten ion ($Z = 74$) with the ionization degree defined from Eqn. (5.40). The bremsstrahlung intensity is calculated from Eqn. (5.41) with the lower limit of integration equal to the frequency. For the indicated temperature and nuclear charge we have: $\bar{q} = 0.241$, $\bar{I} = 775$ eV. Figure 13 ensues that the spectrum of the total emission in both channels exhibits a maximum and a break just at the frequency $\omega = \bar{I}$, which reflects the well-known threshold peculiarities of the recombination emission frequency dependence [9, Chap. 11] (the dominant process in the considered case). At frequencies $\omega < I$, the recombination emission

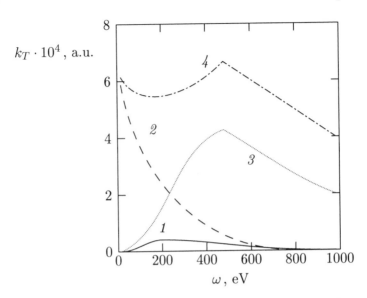

$k_T \cdot 10^4$, a.u.

ω, eV

Fig. 13. Emission spectra of electrons scattering on the tungsten atom ($Z = 74$) averaged over the coronal plasma equilibrium state at temperature $T = 500$ eV. 1 — the polarization bremsstrahlung radiation, 2 — the static bremsstrahlung radiation, 3 — the total polarization bremsstrahlung, 4 — the total static radiation.

corresponds to electron transitions into the states with lower potential energy (the larger main quantum number). In approximation in use these transitions are substituted by a continuous distribution of the target electrons over energy. The relative contribution of polarization channel in the bremsstrahlung and total emission (the R-factor) is virtually the same. The corresponding frequency dependence of the R-factor is a curve that increases as a power law at low frequencies, attains a maximum ($\bar{R}_T^{\max}(T = 500 \text{ eV}, W) \approx 0.6$) at a frequency nearly equal to the temperature, and monotonically decreases at high frequencies due to the effects of the IP penetration into the target core. All the above implies, that in this case the PR fraction is essential in spite of the average charge of the target ion being sufficiently large for the given parameters: $\bar{Z}_i = 18$.

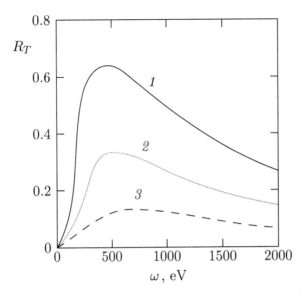

Fig. 14. The R-factor averaged over the coronal equilibrium state at a temperature of 500 eV for different targets: *1* — W, *2* — Mo, *3* — Fe.

Figure 14 demonstrates the spectral dependencies of the R-factor averaged over the coronal equilibrium state of plasma for electron scattering on ions of various elements (Fe, Mo, W) and $T = 500$ eV. It is evident that the PR contribution at the given temperature increases with the nuclear charge: from 0.1 (at the frequency dependence maximum) for iron to 0.6 for tungsten. This increase is due to the temperature-averaged ionization degree of the target decreasing with the nuclear charge growth, which results in increasing the core effective charge that causes the polarization channel radiation. Note that the average ion charge weakly changes here: 16.3 for iron and 18 for tungsten. The R-factor maximum shifts somewhat toward high frequencies in passing from tungsten to iron. The PR contribution in the process decreases with the plasma temperature, increase in linked to the average target ionization degree growth; here the optimal frequency also increases. For example, for $T = 1000$ eV and the tungsten target, when $\bar{q}_T = 0.34$ ($\bar{Z}_i = 25$), the calculation gives $\bar{R}_T^{\max} \approx 0.43$ and $\omega_{\max} \approx 900$ eV.

The analysis of the PR role, carried out within the frames of the suggested method, indicates that the polarization channel contribution can be essential also in the cases of light atoms if the plasma temperature is sufficiently small. For example, for the carbon ions and $T = 10$ eV ($\bar{q}_T = 0.32$) the maximum value of the averaged R-factor is about 0.46. Here the optimal (for the appearance of polarization effects) frequency shifts toward higher frequencies $\omega_{\max} \approx 80$ eV in comparison with the case of heavy elements when $\omega_{\max} \approx T$.

The method developed above allows us to express the photorecombination cross section of a quasi-classical electron with initial energy E through the effective spectral emission as

$$\sigma_r(E, q, Z) = \int_E^{E+I} \frac{d\kappa}{d\omega} \frac{d\omega}{\omega}. \qquad (5.44)$$

In deriving Eqn. (5.44), we have employed the relation between the effective emission and the process cross-section: $\kappa = \omega\sigma$. Equation (5.44) describes both the static and polarization channels if we treat $d\kappa/d\omega$ as the static or polarization effective radiation, respectively.

We have to keep in mind that the classical method employed does not take into account, naturally, virtual excitations of ion's electronic core in a discrete spectrum. Thus, its accuracy depends on the importance of the discrete spectrum contribution in the target polarizability: the smaller this contribution, the higher is the accuracy of the present consideration. The discrete spectrum is insignificant for atoms (ions) with closed electronic shells [26], so the method employed is the most adequate for such targets. In the opposite case, it usually provides, as a rule, the lower bound of the contribution of polarization effects in the considered processes.

The core polarization effects in photorecombination can be characterized by the R-factor in analogy with Eqn. (5.43):

$$R_r(E) = \frac{d\sigma_r^{\text{pol}}(E)}{d\sigma_r^{\text{st}}(E)}. \qquad (5.45)$$

The relative contribution of the polarization channel in the cross-section of the electron recombination on the uranium ion with various ionization degrees q is presented in Fig. 15 as a function of the

IP initial energy. This result is interesting for the interpretation of experiments on the storage rings [66] when the target ionization degree and electron beam energy can be fixed. As in the case of the spectral R-factor for the effective emission (Fig. 12), the role of the ion core polarization in recombination increases with decrease of the quantity q.

For recombination processes in plasma, of interest is the photorecombination rate averaged over the coronal equilibrium state. This can be expressed through the correspondent cross-section as:

$$\alpha_r(T, Z) = 2\sqrt{\frac{2T}{\pi}} \int_0^\infty \sigma_r(xT, \bar{q}(T, Z), Z)e^{-x}x\,dx. \qquad (5.46)$$

Here we have used average ionization degree (5.39) that depends on the temperature and nuclear charge.

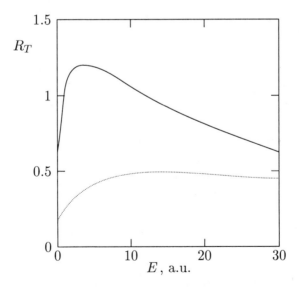

Fig. 15. The R-factor for recombination as a function of the IP initial energy in scattering on the uranium ion with different ionization degree: the solid curve for $q=0.1$, the dashed curve for $q=0.3$.

Specifically, using Eqn. (5.44), in the Kramers approximation
we obtain the photorecombination rate in the static channel:

$$\alpha_r^{\mathrm{Kr}}(T, I) = \left(\frac{2}{3}\right)^{3/2} \frac{8\sqrt{\pi}}{c^3} \frac{1}{\sqrt{T}} \int_0^\infty Z_{\mathrm{ef}}^2(x) e^{-x} \ln\left(1 + \frac{I}{xT}\right) dx.$$
(5.47)

Here $Z_{\mathrm{ef}}(x)$ is the effective ion charge generally depending on the
IP energy. For sufficiently small temperatures, when the penetration
of the recombining electron into the ion core is small, the effective
charge can be set equal to the ion charge and put out from the integral
sign in the right-hand-side of equation (5.47).

Figure 16 shows the photorecombination rates for quasi-classical electrons on the uranium ion in the polarization and static channels as calculated within the frames of the present method. It is
evident that the temperature dependence of the considered photorecombination mechanisms has the different character. The static channel rate increases monotonically. For the polarization channel there
is an optimal temperature value (about 3 a.u. in this case). In in-
terpreting the dependencies calculated from Eqn. (5.46), one should
keep in mind that the average target ionization degree also increases
with temperature. This leads, on the one hand, to the ion effective
charge increase, and to the characteristic frequency growth of the
photorecombination emission, on the other hand . The former increases the rate in the static channel and decreases the rate in the
polarization channel. The second factor increases the polarization
photorecombination rate. As a result, the temperature dependence
of the latter is described by a curve with maximum at which the
polarization channel prevails over the static one for a given element.

The results of the present calculations suggest that there is a
sufficiently broad parameter range where the polarization photorecombination of electrons on multicharged ions is comparable or even
dominates over static recombination. At the same time, one should
keep in mind that the IP energies, for which such domination is effective, are relatively great and comparable with excitation energies of
discrete electron states in the target core. In that case, as a rule, the
dielectronic recombination provides the main contribution to recombination, and the role of the polarization channel could prove essential
in the intermediate energy range. This situation is realized, for ex-

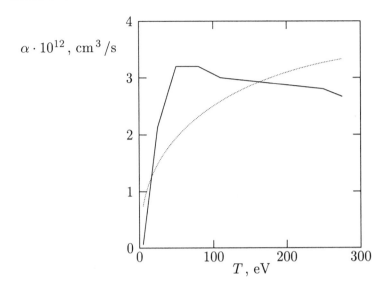

Fig. 16. The recombination rate of the uranium ion with a temperature-dependent ionization degree: the solid curve shows the polarization channel, the dashed curve indicates the static channel.

ample, in the electron – ion recombination of FeII ion, whose rate was calculated in paper [69] by the R-matrix method including 83 states of the electron core of FeIII ion. In [69] the temperature dependence of the process rate was found with taking into account the contribution of photo- and dielectronic recombination in a self-consistent way. It was discovered that there is a broad temperatures range $0.2 - 2$ eV inside which the calculated recombination rate exceeds by several (up to five) times the total contribution from static and dielectronic channels. This is just the case where the polarization recombination mechanism is essential, which was not explicitly distinguished in this paper. Figure 17 presents the results of calculation for the electron – ion recombination rate of FeII in the above temperature range carried out within the frames of the classical method using Eqns (5.44) – (5.46) for the Maxwell electron velocity distribution, together with the results of paper [70]. In this figure also plotted are the temperature dependencies of the photorecombination rate in the static channel calculated in [70] and in the present paper. It is seen that the

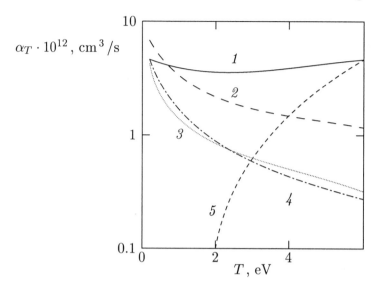

$\alpha_T \cdot 10^{12}$, cm^3/s

Fig. 17. The temperature dependence of the recombination rate of the Fe^{2+} ion:
1 — the total recombination rate as calculated in [69] with account for contributions from 83 states,
2 — the total photorecombination rate as calculated in the present paper assuming the constructive cross-channel interference,
3 — the photorecombination rate calculated in the Kramers approximation according to Eqn. (5.47),
4 — the static photorecombination rate from paper [70],
5 — the dielectronic recombination rate from paper [71].

simple model employed here allows us to reproduce (with reasonable accuracy) the results of very awkward calculations [69] and give them a visual physical interpretation. Besides, from comparison of curves describing static photorecombination, we find good accuracy of the Kramers approximation in the considered temperature range.

Thus, in the present section within the frames of the plasma model for the electronic core of ion and RA for IP, we have analyzed the role of PR both for non-averaged and averaged over the coronal equilibrium state emission spectra and photorecombination rate in dependence of the problem parameters. It has been shown that the polarization channel contribution can be comparable to and can even exceed the ordinary (static) channel contribution, including the case of multicharged ions, provided that the number of bound electrons in the core is sufficiently large. For a given fixed temperature, the role of the polarization effects in emission increases with the growth of the target nuclear charge owing to the electronic core effective polarization charge increase. With increasing temperature the ion ionization degree increases and the relative PR intensity decreases.

It has been demonstrated that the polarization mechanism contribution to the photorecombination rate can prevail over the static contribution in the case of the sufficiently heavy ions in the certain interval of energies and temperatures.

In conclusion we note that the results of the present consideration should be taken into account, in particular, for correction of plasma diagnostic methods. Indeed, the account of the polarization channel changes the relation between the intensity of the experimentally observed continuum spectrum of emission and the average ion charge. This relationship, which had been traditionally calculated in the static approximation, should be added with the correction multiplier $1 + R$. As a result, the real average ion charge turns out to be less, generally speaking, than its value as derived in the static approximation.

5.3. The transition bremsstrahlung of thermal electrons on plasma ions

The transition BR on ions in plasma is a special case of the polarization BR when the Debye "coat" around charged particles in plasma serves as an atom–target. For fast IP this problem is considered in Sect. 6 of the present review. Here we consider the case for incident electrons with thermal energies when the Born approximation for BR can not be apply [72]. Then, however, the quasi-classical method can be employed. The dipole moment induced in the IP target core in this approximation reads

$$\mathbf{D}_{\mathrm{pol}}(\mathbf{R}, \omega) = -\frac{\mathbf{R}}{R^3} \int_0^R \beta(r, \omega) 4\pi r^2 \, dr \,. \qquad (5.48)$$

Here $\mathbf{R}$ is the IP radius-vector, $\beta(r, \omega)$ is the dynamic polarizability spatial density. Dipole moment (5.48) determines the spectral effective emission in the polarization channel in accordance with the formula:

$$\frac{d\kappa_{\text{pol}}(\omega)}{d\omega} = \frac{4\omega^4}{3c^3} \int_0^\infty |\mathbf{D}_{\text{pol}}^\omega(\omega, \rho)|^2 \rho \, d\rho. \tag{5.49}$$

Here $\mathbf{D}_{\text{pol}}^\omega(\omega, \rho)$ is the Fourier image of polarization dipole moment (5.48) at frequency ω calculated along the incident particle trajectory characterized by the impact parameter ρ. In the case of the Debye sphere, for the polarizability spatial density the high frequency approximation is valid:

$$\beta_\infty(\omega) = -\frac{\omega_{\text{pe}}^2(r)}{4\pi\omega^2} = -\frac{n_e(r)}{\omega^2} \tag{5.50}$$

unless the emission frequency ω is too close to the average electron plasma frequency $\bar{\omega}_{\text{pe}} = \sqrt{4\pi\bar{n}_e}$ ($\bar{n}_e$ is the average number density of plasma electrons). By expressing the electron number density inside the Debye sphere in the form

$$n_e(r) = \frac{Z_i}{4\pi r_{\text{De}}^2} \frac{\exp(-r/r_{\text{De}})}{r} \tag{5.51}$$

where r_{De} is the electron Debye radius, for the induced dipole moment we obtain from Eqn. (5.48):

$$\mathbf{D}_{\text{pol}}^{(\text{D})}(\omega, \mathbf{R}) = \frac{1}{\omega^2} \frac{\mathbf{R}}{R^3} N_e(R). \tag{5.52}$$

Here $N_e(R)$ is the number of plasma electrons inside the sphere of radius R which is given by the expression:

$$N_e(R) = \int_0^R 4\pi n_e(r) r^2 \, dr = Z_i \left[1 - e^{-R/r_{\text{De}}}(1 + R/r_{\text{De}}) \right]. \tag{5.53}$$

To calculate the total effective (summed over the impact parameter and frequency) emission of electron with energy E in polarization channel on the Debye sphere (or the total bremsstrahlung losses), we proceed from the expression:

$$\kappa_{\text{pol}} = \int_{\omega_{\text{pe}}}^{E/\hbar} d\kappa_{\text{pol}}(\omega). \tag{5.54}$$

Substituting Eqns (5.48) and (5.49) into it we have:

$$\kappa_{\text{pol}}^{(D)}(E) = \frac{4}{3c^3} \int_0^\infty \rho \, d\rho \iint dt \, dt' \int_0^{E/\hbar} e^{i\omega(t-t')}$$

$$\times \frac{\mathbf{R}(t)\mathbf{R}(t')}{R^3(t)R^3(t')} N(R(t))N(R(t')) \, d\omega \,. \qquad (5.55)$$

Here the lower limit of integration over frequency is set at zero. In reality, transversal photons are known to propagate in plasma if the condition $\omega > \omega_{\text{pe}}$ is fulfilled. The above substitution is adequate to the ideal plasma case when the characteristic correlation time in the scattering electrons motion is less than the inverse plasma frequency.

Next, we use the equality

$$\int_0^\infty e^{i\omega(t-t')} \, d\omega = \pi\delta(t - t') \qquad (5.56)$$

(the upper limit is set infinity in accordance with the quasi-classical condition $\hbar \to 0$). We employ Eqn. (5.55), pass to the integration variable R (after such a substitution the lower limit of integration becomes equal to the minimal distance between the IP and the ion $r_{\min}(\rho)$ and the result increases by two times due to the sub-integral function in Eqn. (5.55) being even relative to the change of the time sign). After integrating over the impact parameter ρ, as has been shown in [45], we find (here and below we use atomic units):

$$\kappa_{\text{pol}}^{(D)} = \frac{8\pi}{3c^3\sqrt{2T}} \int_0^\infty f_{\text{pol}}^2(r) \sqrt{1 - \frac{U_D(r)}{T}} \, r^2 \, dr \,. \qquad (5.57)$$

Here $U_D(r) = -Z_i \exp(-r/r_{\text{De}})/r$ is the screened (Debye) potential of an ion in plasma, $f_{\text{pol}}(r)$ is the "polarization" force defined by the expression:

$$f_{\text{pol}}(r) = \frac{N_e(r)}{r^2} \,. \qquad (5.58)$$

This force (of repulsion) acts on IP from the target electrons placed inside the sphere of radius R. With the same force (in accordance with the 3rd Newton's law) IP accelerates the target electrons, which move as a single negative charge cloud, thus producing BR.

We also provide here the expression for the total effective emission in the static channel [25]:

$$\kappa_{\rm st}^{(D)} = \frac{8\pi}{3c^3\sqrt{2T}} \int_{r_{\rm min}}^{\infty} f_{\rm st}^2(r) \sqrt{1 - \frac{U_{\rm D}(r)}{T}}\, r^2 dr. \tag{5.59}$$

Here $f_{\rm st}(r) = -dU_{\rm D}/dr$ is the ordinary "static" force that defines the IP trajectory.

Note that in spite of a large similarity between Eqns (5.57) and (5.59), there is an essential difference between them: integral (5.59) diverges in the lower limit (in the quasi-classical case as $\int_0 r^{-5/2} dr$), while the integral in Eqn. (5.57) converges in the lower limit, the latter being because the plasma electron charge $N_{\rm e}(R)$ that radiates by the polarization channel tends to zero as R decreases, as follows from Eqn. (5.53).

Employing Eqns (5.53) and (5.57) and the expression for the Debye potential, we obtain the total effective emission in polarization channel in the form:

$$\kappa_{\rm pol}^{(D)} = \frac{8\pi}{3c^3\sqrt{2T}} \frac{Z_{\rm i}^2}{r_{\rm De}} \Phi\left(\frac{2a_T}{r_{\rm De}}\right). \tag{5.60}$$

Here we have introduced the function:

$$\Phi(x) = \int_0^{\infty} [1 - (1+r)e^{-r}]^2 \sqrt{1 + (x/r)e^{-x}}\, dr/r^2 \tag{5.61}$$

The parameter $a_T = Z_{\rm i}/2T$ is the Coulomb scattering length of electron of energy T on an ion with charge $Z_{\rm i}$. Note that the ratio $2a_T/r_{\rm De}$ is inversely proportional to the plasma non-ideality parameter. For the ideal plasma $2a_T/r_{\rm De} \ll 1$. The function $\Phi(x)$ slowly increases with increase of the argument, so that for the ideal plasma we can put $\Phi(\xi) = 0.5$. Then expression (5.60) takes the form:

$$\kappa_{\rm pol}^{(D)} = \frac{4\pi}{3c^3\sqrt{2T}} \frac{Z_{\rm i}^2}{r_{\rm De}}. \tag{5.62}$$

Equation (5.62), which is valid for the quasi-classical IP movement, coincides (per one ion) with the expression for the total power emitted in the polarization channel obtained in [73] for the straight-flying over-thermal electrons.

Thus we can conclude that under the ideal plasma conditions, the character of the IP movement weakly affects its total effective radiation in the polarization channel.

For the effective static BR in the Coulomb field of an ion, considering the cutting off of the integral at the lower limit, we obtain from Eqn. (5.59):

$$\kappa_{st} = \frac{8\pi Z_i \sqrt{2T}}{9c^3} \left[\left(\frac{x_m + 2}{x_m} \right)^{3/2} - 1 \right] . \qquad (5.63)$$

Here $x_m = (2\sqrt{2T}/Z_i)^{2/3}$.

In the limit $x_m \ll 1$ equation (5.63) simplifies to the form:

$$\kappa_{st} = \frac{8\sqrt{2}\pi}{9c^3} Z_i^2 . \qquad (5.64)$$

Result (5.64) differs from the Kramers formula only by the numerical coefficient about 0.8, which is explained by using the cut-off radius with an approximate numerical factor.

From equations (5.62) and (5.64) and using the expression for the Debye radius, we obtain the ratio of polarization and static channel contributions:

$$\Re^D(n_e, T) = \frac{\kappa_{pol}^D}{\kappa_{st}} \cong 3 \frac{\sqrt{n_e}}{T} . \qquad (5.65)$$

As is evident from Eqn. (5.65), plasma must be as cold and dense as possible for the contribution of polarization effects into BR on an ion with the Debye screening to be appreciable.

Let us evaluate numerically the quantity $\Re$ for a laser plasma with the following parameters: $n_e \approx 7 \cdot 10^{18}$ cm^{-3}, $T \approx 1$ eV, then $\Re \approx 10\%$. If $n_e \approx 7 \cdot 10^{20}$ cm^{-3}, then $\Re \approx 100\%$, but then the plasma parameter becomes less than unity and plasma turns non-ideal. It is interesting to evaluate the contribution of the polarization BR for plasma of the inner regions of the Sun: $n_e \approx 5.7 \cdot 10^{25}$ cm^{-3}, $T \approx 1550$ eV. For these values with Eqn. (5.65) we find: $\Re \approx 15\%$.

Let us rewrite the expression for the ratio $\Re$ through the plasma ideality parameter $\zeta = (4\pi/3)r_{\mathrm{D}}^3 n_{\mathrm{e}}$:

$$\Re(n, \zeta) \approx 1.24 \frac{n^{1/6}}{\zeta^{2/3}}. \tag{5.66}$$

As is evident from this formula, for a fixed plasma ideality parameter ζ the ratio $\Re$ is weakly depending function on the plasma electron number density.

The above consideration entails that the polarization contribution to the total bremsstrahlung losses of plasma electrons on the Debye cloud around an ion in a non-degenerate plasma can be comparable with the contribution due to the ordinary (static) BR only in the case of the sufficiently cold and dense plasma, when the parameter of ideality ζ is about unity. Otherwise, the ratio of contributions due to polarization and static channels does not exceeds $10-15\%$.

The spectral R-factor in the RA frames can be obtained using Eqns (5.8) and (5.9) and the corresponding expressions for the potential and electron density of the Debye sphere:

$$R^{(\mathrm{rot})}(\omega) = \left[1 - \frac{\exp\left(r_\omega/r_{\mathrm{De}}\right)}{1 + r_\omega/r_{\mathrm{De}}}\right]^2 \approx (r_\omega < r_{\mathrm{De}}) \approx \frac{1}{4}\left(\frac{r_\omega}{r_{\mathrm{De}}}\right)^4. \tag{5.67}$$

Here $r_\omega = r_{\mathrm{eff}}(\omega)$ is the solution to equation (5.10). The inequality $r_\omega < r_{\mathrm{De}}$ employed in Eqn. (5.67) is correct for the ideal plasma in the frequency range $\omega \approx \omega_{\mathrm{C}} = v^3/Z_{\mathrm{i}}$, i.e. for frequencies of the order (and above) the typical Coulomb frequency. Equation (5.67) implies that the polarization channel contribution to the spectral BR cross-section for thermal energy electrons on the Debye sphere at frequencies $\omega > \omega_{\mathrm{pe}}$, calculated within the rotation approximation frame, is small. The estimate shows that in this case $R^{(\mathrm{rot})}(\omega) \leq 1\%$.

The law of $R^{(\mathrm{rot})}(\omega)$ decreasing with frequency can be explicitly obtained by employing the expression for r_ω in the Coulomb field: $r_\omega^{\mathrm{C}} \approx \sqrt[3]{2 Z_{\mathrm{i}}/\omega^2}$. Then from Eqn. (5.67) we find:

$$R_{\mathrm{D}}^{\mathrm{rot}}(\omega) \approx \frac{1}{4 r_{\mathrm{D}}^4} \frac{(2 Z_{\mathrm{i}})^{4/3}}{\omega^{8/3}}. \tag{5.68}$$

Since the spectral effective emission in static channel weakly depends on frequency in the considered range, Eqn. (5.68) yields the frequency dependence of the spectral effective emission of thermal electrons on the Debye sphere in polarization channel in the rotation approxima- tion:

$$\frac{d\kappa_{\text{pol}}^{(\text{rot})}(\omega)}{d\omega} \propto \omega^{-\frac{8}{3}}. \tag{5.69}$$

Dependence (5.69) somewhat differs from the decreasing cross section of the transition (polarization) BR of fast over-thermal elec- trons in the frequency range $\omega > (v/v_{Te})\,\omega_{\text{pe}}$, according to which

$$d\kappa_{\text{pol}}^{(\text{rot})}(\omega)/d\omega \propto \omega^{-4}. \tag{5.70}$$

5.4. Quantum calculation (by the incident particle motion) of the effective radiation on multielectron ions

In this section, the plasma model for the ion – target is em- ployed to calculate the polarization BR in quantum description of the motion of IP scattering on ion.

Quantum calculation of BR in statistic channel for the Thomas – Fermi ion was first carried out in papers [55, 56]. This cal- culation demonstrated a good accordance with consistent quantum- mechanical calculations for the Hartree – Fock target core and, more- over, confirmed a high accuracy of the rotation approximation of the Kramers electrodynamics [47].

The operator of the IP motion perturbation, producing BR in polarization channel, can be recovered from expression (5.47) for the induced dipole moment in the ion core. The corresponding formula has the form:

$$V_{\text{pol}}(\mathbf{R}, \omega) = \frac{\mathbf{R}\,\mathbf{E}(\omega)}{R^3} \int_0^R \beta(r, \omega)\, 4\pi r^2\, dr. \tag{5.71}$$

Here $\mathbf{E}(\omega)$ is the electric vector in the radiation field. Essentially, this formula describes the non-dipole potential of the IP interaction with the perturbed ion core, which is expressed in the presence of the IP radius – vector R module in the upper limit of integration. This fact has a simple electrostatic interpretation: the external charge

interacts only with a part of the electronic cloud inside the sphere of
radius R if the process occurs without exciting bound electrons of
the target.

Quantum calculation of the radiation intensity is significantly
simplified due to spherical symmetry of the scattering potential. In
this case the standard method [2] of the IP wave function expansion in
spherical harmonics or in the quantum number of the orbital moment
l can be applied. The component of wave function corresponding to
a fixed value of l is the product of the radial and angular parts.
The angular part is known to be the spherical function. The radial
part $u(r, l, p)$ satisfies the Schrödinger equation with the following
boundary condition at infinity:

$$u(r \to \infty, l, p) \to \frac{2}{r} \sin \left(pr + \frac{Z_i}{p} \ln(2pr) - \frac{\pi}{2} l + \delta(l, p) \right) . \quad (5.72)$$

Here p is the IP momentum, $\delta(l, p) = \delta^C(l, p) + \Delta\delta(l, p)$ is the total phase shift equal to the sum of the Coulomb $\delta^C(l, p)$ and the
non-Coulomb $\Delta\delta(l, p)$ phase shifts which can be calculated from the
formula [74]:

$$\sin(\Delta\delta(l, p)) = \frac{1}{2p} \int_0^\infty \left(\frac{Z_i}{r} - |U(r)| r \right) u(r, l, p) \, u^{\text{Coul}}(r, l, p) \, r^2 \, dr .$$

$$(5.73)$$

Here $u^{\text{Coul}}(r, l, p)$ is the solution of the Schrödinger radial equation
with the Coulomb ion potential, $U(r)$ is the IP potential in the ion –
target field (5.37).

For numerical calculations it is convenient to use the auxiliary
radial wave function: $v(r, l, p) = r^{-l} u(r, l, p)$. The corresponding
Schrödinger equation reads:

$$v'' + 2 \frac{l+1}{r} v' + \left(p^2 - 2U(r) \right) v = 0 \quad (5.74)$$

(the prime denotes differentiation with respect to radius) subjected
to following boundary conditions:

$$v(0) = 1, \quad v'(0) = -\frac{Z}{l+1} . \quad (5.75)$$

Here Z is the ion nuclear charge.

In order to satisfy asymptotic (5.72), we introduce the normalizing coefficient N:

$$N = \left\{ r^{l+1} \sqrt{v^2 + [p^{-1} v']^2} \right\}_{r \to \infty}. \qquad (5.76)$$

Finally, we have for the wave function:

$$u(r, l, p) = \left(\frac{2}{N} \right) r^{l+1} v(r, l, p). \qquad (5.77)$$

With the use of function (5.77) for the BR spectral intensity we arrive at the following expression:

$$\frac{dW}{d\omega} = \frac{2}{3 \, c^3 \, p_i^3 \, p_f} \sum_{l=0}^{\infty} (l + 1) \left[|M_{l,l+1}|^2 + |M_{l+1,l}|^2 \right]. \qquad (5.78)$$

Here we have introduced the radial matrix elements $M_{l,l+1}, M_{l+1,l}$ taken between wave functions (5.77) of the module of the force effecting on IP and producing BR in the certain channel. Moreover, for the static channel the expression for the corresponding force is given by Eqn. (5.2) — this is an ordinary force determining the IP motion in the static field of the ion – target.

The absolute value of the force that produces radiation in polarization channel can be recovered from Eqn. (5.71). It is determined by the non-dipole dynamic polarizability of the ion core and reads

$$f^{\text{pol}} = -\frac{\omega^2}{R^2} \int_0^R \beta(r, \omega) \, 4\pi \, r^2 \, dr. \qquad (5.79)$$

Since the expression for the spatial density of polarizability has, generally speaking, an imaginary part, polarization force (5.79) and the corresponding matrix element $M_{l,l'}$ have the imaginary components together with the real ones. The radial matrix element of static force (5.2) is, naturally, purely real.

The total matrix element in Eqn. (5.78) is the sum of the static and polarization terms. Their real parts produce the interference term in the expression for the BR intensity, and the imaginary part of the polarization matrix element provides no contribution to the cross-channel interference.

Formally, the infinitive series in the quantum number of the orbital momentum l in expression (5.78) for the BR intensity rapidly converges for strongly inelastic processes. For example, for $\omega/E \approx$ $0.7 - 0.9$ (in the IP scattering on the KII ion) the main contribution to the BR intensity is provided by the first $3 - 4$ terms of this series. Note that the situation is quite opposite for weakly inelastic processes when the series in l converges extremely slowly.

The analysis of the calculated data indicates that at low and high frequencies the main contribution to the polarization BR is due to the real part of the induced dipole moment in the target core. At the "moderate" frequency range ($I_p < \omega \leq Z$), the imaginary part of the core dipole moment dominates in the polarization channel. This conclusion also follows from the PBR calculations within the frames of the random phases approximation with changes for the core polarizability [30].

The results of quantum calculations of the spectral R-factor and the relative contributions of the interference term into the BR intensity in scattering of IP on the KII ion are shown in Fig. 18 for two values of the ratio ω/E (Fig. 18(a) and (b), respectively). The calculation has been performed assuming the local plasma density approximation and using the ion core electron density as in the Thomas – Fermi – Dirac model. Here it should be noted that the statistic approximation is known to describe well an atom's (ion's) properties in the localization region of most its electrons. In the near-nuclear region and at the ion's boundary, where one-electron effects become significant, the accuracy of the statistical approximation noticeably worsens. In particular, in the statistic model frames, at the ion's boundary the electron density and local plasma frequency vanish. Here single-electron excitations significantly contribute to the polarizability. They should be taken into account at frequencies of the order of the core ionization potential, where the size of electronic orbit gets larger than the ionic core size in the statistical model. In fact, here in addition to the collective plasma frequency of electron density oscillations, proper electron oscillations in the core field are manifest. This effect can be approximately taken into account by shifting the frequency ω in the formula for the target polarizability by the amount $\Delta\omega$ such that the photoabsorption maximum fall at the single-electron ionization potential.

Figure 18 demonstrates that the spectral R-factor has a maximum near the target ionization potential, whose width decreases with the decrease in the inelasticity degree of the process (the ratio ω/E); at the same time the maximal value of the R-factor somewhat increases.

The main difference from the results obtained in the rotation approximation is that the spectral R-factor calculated for the IP quantum motion decreases more sharply with frequency in the "moderate" frequency range $I_p < \omega \leq Z$.

Figure 18(b) demonstrates the value and character of the cross-channel interference as a function of the BR frequency. At frequencies $\omega \leq I_p$ the interference has destructive character (decreases the total intensity of the process) and has an appreciable value. At $I_p < \omega \leq Z$ the interference term changes the sign and increases the total intensity (constructive interference) by staying quite large. The cross-channel interference is negligible at low frequencies. At high frequencies its contribution is about $10-20\%$, decreasing as frequency increase.

As the inelasticity degree of BR gets smaller, the role of the interference diminishes because the spatial formation regions of the static and polarization radiation overlap smaller. This is seen from Fig. 18(b), which also suggests that for a less inelastic process the frequency range of the destructive interference proves somewhat extended toward high frequencies. For $\omega \leq I_p$ the character and amplitude of the cross-channel interference weakly depend on the degree of inelasticity of the process.

Thus, the analysis performed in this Section for polarization and interference effects in a strongly inelastic BR on an ion for a quantum IP motion, while demonstrating qualitative adequacy of the generalized rotation approximation developed earlier, has notably corrected this approximation in the quantitative aspect by completing the general phenomenological picture with the cross-interference studies. The main conclusion from the quantum treatment is that the polarization channel contribution to the strongly inelastic BR intensity increases from zero as a power law, is mostly significant near the ionization frequency of the ion-target, and drastically falls off

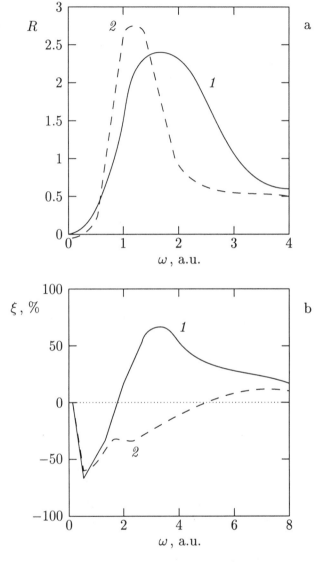

Fig. 18. The results of quantum (on the IP motion) calculation of the spectral R-factor (a) and the channel interference contribution ξ to the BR intensity (b) performed in the frames of the static model for two values of the inelasticity parameter: $1 — \omega/E = 0.9$, $2 — \omega/E = 0.6$ in electron scattering on the ion KII [25].

with further frequency growth due to IP penetration into the target's core. Here the width of the frequency maximum for the R-factor increases with the process inelasticity.

6. Polarization channels of fast particle radiation on atoms, in plasma, and in a dense medium

6.1. Polarization bremsstrahlung radiation of a fast charged particle on a Thomas–Fermi atom

The PBR spectral cross sections (3.40) and (3.41) obtained in Sect. 3 are valid for $\omega \gg I$ where I is the characteristic ionization potential of the atom. In the case of multielectron atoms, this quantity is rather uncertain so that the applicability range of the high frequency approximation must be specified.

At the same time, polarization effects in BR should be most substantial exactly for multielectron atoms. Calculation of the dynamic polarizability of a multielectron atom that determines the PBR cross section is a difficult quantum mechanical task, which has to be solved anew for each particular target. In this connection, utilizing simple universal models relevant for the assessment of the polarization BR cross section and revealing general qualitative relationships of this process seems to be useful. One such model is the local electron density (or local plasma density) method first suggested by Brandt and Lundqvist to calculate the photoabsorption cross section by multielectron atoms [3] (see Sect. 2.1).

The spectral cross section of polarization BR of an electron on atom within the frames of the first Born approximation is described by Eqn. (3.33), which can be simplified for the process without target excitation, as was shown in the previous Section (here and below we use relativistic units with $\hbar = c = 1$):

$$\frac{d\sigma^{\mathrm{PB}}}{d\omega} = \frac{\omega^5}{(2\pi)^3 V} \int d\Omega_{\mathbf{n}}\, d\mathbf{q}\, |\alpha(\omega, \mathbf{n} + \mathbf{k})|^2 \left[\mathbf{n}\mathbf{A}(q)\right]^2 \delta(\omega + \mathbf{q}\mathbf{v}). \quad (6.1)$$

Here $d\Omega_{\mathbf{n}}$ is the solid angle of the photon emission direction, $\mathbf{k}, \omega$ the wave vector and the bremsstrahlung photon frequency, respectively,

$\mathbf{q} = \mathbf{p}_f - \mathbf{p}_i$ is the change in the incident particle momentum, and $\mathbf{A}(q)$ vector – potential of the incident particle's field in the momentum representation which in the axial gauge ($A_0 = 0$) is given by Eqn. (3.29). For further calculation of the BR cross section we restrict ourselves to the Born – Bethe approximation in which we can put

$$\alpha(\omega, q) = \alpha(\omega)\theta(p_a - q) \qquad (6.2)$$

where $\theta(x)$ is the unitary step function. The characteristic atomic momentum will be taken in the Thomas – Fermi form $p_a = Z^{1/3}/(ba_0)$. Then from (6.1) and (6.2) we have:

$$\frac{d\sigma^{PB}}{d\omega} = \frac{4\omega^3}{V^2}|\alpha(\omega)|^2 \left\{ \theta\left(\frac{p_a V}{1+V} - \omega\right) [H_1(\omega, p_a - \omega) + H_2(\omega)] \right.$$
$$\left. + \theta\left(\omega - \frac{p_a V}{1+V}\right) H_1\left(\omega, \frac{\omega}{V}\right) \right\}, \qquad (6.3)$$

$$H_1(\omega, q_{min}) = \int_{q_{min}}^{p_a} G_1(q, \omega) \frac{q \, dq}{(q^2 - \omega^2)^2},$$

$$H_2(\omega) = \int_{\omega/V}^{p_a - \omega} G_2(q, \omega) \frac{q \, dq}{(q^2 - \omega^2)^2}.$$

Here

$$G_1 = \frac{p_a^2 - (q-\omega)^2}{2\omega q}\left[\omega^2 V^2 + q^2 - \frac{5}{2}\omega^2 + \frac{\omega^4}{2q^2 V^2}\right]$$
$$- \frac{1}{3}\left[\left(\frac{p_a^2 - (q-\omega)^2}{2\omega q}\right)^3 + 1\right]\left[q^2 - \frac{5}{2}\omega^2 + \frac{3\omega^4}{2q^2 V^2}\right], \qquad (6.4)$$

$$G_2 = 2\omega^2\left(V^2 - \frac{5}{3}\right) = \frac{4}{3}q^2.$$

At low frequencies $\omega < p_a V$, when the first term in curly brackets contributes mostly to the cross section, Eqn. (6.3) can be reduced to the known expression for the polarization BR spectral cross section of a relativistic incident electron [61] (see also Eqn. (3.38) for the spectral-angular PBR cross section):

$$\frac{d\sigma^{\mathrm{PB}}}{d\omega} = \frac{16\omega^3|\alpha(\omega)|^2}{3V^2} \ln\left(\frac{2\gamma p_{\mathrm{a}}V}{\omega(1+V)}\right), \quad \omega < p_{\mathrm{a}}V. \tag{6.5}$$

Here $\gamma = (1 - V^2)^{-1/2}$ is the relativistic Lorentz-factor, $\alpha(\omega)$ dipole dynamic polarizability of the atom – target.

Passing in Eqn. (6.5) to dimensionless variables, we get the following expression for the spectral cross section of polarization bremsstrahlung radiation:

$$\begin{aligned}
d\sigma^{\mathrm{PB}}(\nu) &= \frac{16Z^2 b^6}{3V^2} |\nu^2\beta(\nu)|^2 \frac{d\nu}{\nu} \ln\left(\frac{2\gamma V}{\nu a_0(1+V)Z^{2/3}}\right) \\
&= Z^2 d\tilde{\sigma}^{\mathrm{PB}}(\nu).
\end{aligned} \tag{6.6}$$

Here the function $d\tilde{\sigma}^{\mathrm{PB}}(\nu)$ is introduced which can be naturally called the reduced cross section of the process. It reveals an approximate scaling over the ω/Z parameter, while the remaining dependence on the charge has logarithmic character.

The spectral cross section of ordinary (static) BR with allowance for the nuclear field screening in the case of inelastic electron scattering is given by the expression

$$d\sigma^{\mathrm{SB}}(\omega) = \frac{16Z^2}{3V^2} \frac{d\omega}{\omega} \ln\left\{\frac{V}{p_{\mathrm{a}}}\right\}, \quad \omega < p_{\mathrm{a}}V. \tag{6.7}$$

The cross section ratio of polarization and static BR (R-factor) at frequencies $\omega < p_{\mathrm{a}}V$ for relativistic particles is

$$R(\nu, Z, \gamma) \equiv \frac{d\sigma^{\mathrm{PB}}}{d\sigma^{\mathrm{SB}}} = b^6|\nu^2\beta(\nu)|^2 \frac{\ln\left\{\frac{137\gamma}{\nu Z^{2/3}}\right\}}{\ln\left\{\frac{137}{Z^{1/3}}\right\}}, \quad \nu < \frac{137}{Z^{2/3}}. \tag{6.8}$$

R-factors calculated as a function of dimensionless frequency ν for different γ and $\nu < 137/Z^{2/3}$ are shown in Fig. 19.

Analysis of general expression (6.1) indicates that at high frequencies $\omega > p_{\mathrm{a}}V$ the emission diagram of the polarization channel narrows, so the effective emission angles satisfy the inequality

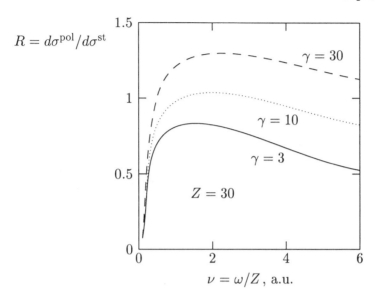

Fig. 19. The R-factor dependence on the reduced frequency ν calculated for fast electron BR on a Thomas–Fermi atom with nuclear charge $Z = 30$ at different values of the relativistic Lorentz-factor $\gamma = 3, 10, 30$.

$\vartheta \geq \sqrt{p_a/\omega}$. Within the frequency range $p_a < \omega < \gamma^2 p_a$ there are the BR angles $\gamma^{-1}, \vartheta \geq \sqrt{p_a/\omega}$ where a polarization mechanism dominates over the ordinary (static) radiative mechanism. The PBR emission beam getting narrow is presented in Fig. 20.

As seen from Eqn. (6.6), dynamical properties of a statistical atom within the Born–Bethe approximation are determined by the function $g(\nu) = |\beta(\nu)/\beta^{\mathrm{hf}}(\nu)|$, which can be approximated (within the $\pm 5\%$-error) as follows:

$$g_{\mathrm{fit}}(\nu) = \left(1 - e^{-2.2\nu}\right)^{0.7} \tag{6.9}$$

Within the Born–Bethe approximation it is easy to obtain a simple analytical expression for fast electron bremsstrahlung losses on statistical atom via polarization channel within the given spectral range $0 < \nu < \nu_{\mathrm{h}}$. To this aim, Eqn. (6.9) can be simplified to the form

$$g_{\mathrm{a}}(\nu) = \sqrt{(\nu - 1)\theta(1 - \nu) + 1} \tag{6.10}$$

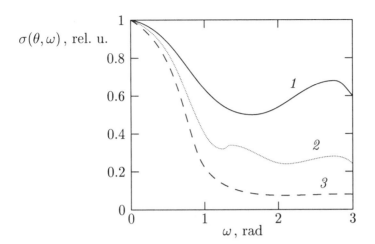

Fig. 20. The PBR angular dependence on an atom ($Z = 30$) for different bremsstrahlung photon energies $\omega = 5$ keV (curve *1*), $\omega = 15$ keV (curve *2*), $\omega = 50$ keV (curve *3*); the IP velocity is $V = 0.9c$.

Substituting Eqn. (6.10) into the formula for the spectral cross section and integrating over frequency the expression for spectral losses within the limits specified, we find

$$W_{\text{pol}}^{\text{B}-\text{B}}(\nu_{\text{h}}) \simeq \frac{16Z^3}{3c^3V^2}\left[\nu_{\text{h}}(1 + \ln(\nu_{\text{max}}/n u_{\text{h}})) - \ln\sqrt{\nu_{\text{max}}} - 3/4\right],$$

$$1 < \nu_{\text{h}} < \nu_{\text{max}}. \tag{6.11}$$

Here $\nu_{\text{max}} = \gamma V/(bZ^{2/3})$.

Polarization BR effects are large and their description relatively simple in the case of fast electron scattering on negative ions. This problem was studied in [52] for the negative hydrogen ion. The following spectral cross section was obtained within the Born – Bethe approximation:

$$d\sigma(\omega) = \frac{4V_{\text{i}}}{3\pi c^3}\frac{d\omega}{\omega}\sigma_{\text{tr}} + \frac{16d\omega}{3c^3V_{\text{i}}^2\omega}[1 + \omega^2\alpha(\omega)]^2\ln\left(\frac{\lambda V_{\text{i}}}{R_0\omega}\right), \quad (6.12)$$

where σ_{tr} is the electron scattering transport cross section on neutral atom, λ is an uncertain coefficient of the order unity. In equation (6.12) the first term comes from scattering on a neutral atom and the second from scattering on an weakly coupled electron, with the first term in the square brackets describing the static channel and the second term the polarization channel. The frequency dependence of the total BR cross section calculated using Eqn. (6.12) represents a curve with maximum at which the cross section takes the value almost two times as large as that obtained ignoring the polarization term.

6.2. Fast charged particle polarization BR cross section on ions in a plasma

Consider first bremsstrahlung radiation caused by the conversion of a virtual photon of the incident particle's field into a real one on plasma electrons. Its cross section is given by the term in Eqn. (3.66) proportional to the electron's dynamic form factor. The corresponding expression with account of the explicit form of $S^{(e)}(q)$ (3.75) reads

$$
d\sigma^{\mathrm{pl}}(\mathbf{k}, \mathbf{q}) = \frac{1}{V_0} \left(\frac{e^2}{m} \right)^2 \left[\mathbf{n}\mathbf{A}^{(0)}(q) \right]^2 \left\{ |\delta n_e|_q^2 \left| \frac{\tilde{\epsilon}_q^{l(i)}}{\tilde{\epsilon}_q^l} \right|^2 \right.
$$

$$
\left. + |\delta n_i|_q^2 \left| \frac{1 - \tilde{\epsilon}_q^{l(e)}}{\tilde{\epsilon}_q^l} \right|^2 \right\} \frac{\omega \, d\omega \, d\Omega_{\mathbf{k}} \, d\mathbf{q}}{(2\pi)^4}. \qquad (6.13)
$$

The plasma component density fluctuations entering this expression are given by the 2-d formula from (3.77).

The terms in the curly brackets in Eqn. (6.13) describe the process of virtual photon scattering into real photon on plasma electrons. The first of them comes from the energy – momentum q transmission to a subsystem of plasma electrons. The second describes the q 4-vector transmission to plasma ions via the IP field interaction with plasma electrons, which is the consequence of the Coulomb interaction between electron and ion plasma components (Debye's screening).

Consider first the fraction of cross section (6.13) determined by the second term in the curly brackets. It coincides with the polarization (transition) BR cross section in collision of a fast charged particle with a point-like plasma ion at frequencies $\omega \gg \omega_{\mathrm{pe}}$ considered in papers [73, 75]. To make sure of this, note that for an ion at rest and ignoring recoil the following limiting transition holds:

$$\lim_{v_{\text{Ti}} \to 0} \frac{1}{\sqrt{2\pi}|\mathbf{q}|V_{\text{Ti}}} \exp\left[-\frac{(q^0)^2}{2\mathbf{q}^2 V_{\text{Ti}}^2}\right] = \delta(q^0). \tag{6.14}$$

Here the situation is considered when the inequalities hold: $q^0 \gg |\mathbf{q}|V_{\text{Te}}$, $\omega \gg \omega_{\text{pe}}$. In this case, virtual photon scattering on the electronic polarization charge around IP can be disregarded. Indeed, this charge is proportional to the small ratio $\left|(1 - \tilde{\epsilon}_q^{l(e)})/\tilde{\epsilon}_q^l\right| \approx \frac{\omega_{\text{pe}}^2}{\omega^2} \ll 1$. So PBR is caused solely by the proper IP field scattering on the electronic charge around ion, which is taken into account in Eqn. (6.13).

Considering Eqn. (6.14) and the explicit form of the IP electromagnetic potential (see Eqns (6.5) and (3.29)), we find the PBR cross section on the Debye sphere around ion in a plasma:

$$d\sigma^{\text{pl}}(\mathbf{k}, \mathbf{q}) = \frac{1}{V_0}\left(\frac{e^2}{m}\right)^2 \frac{[\mathbf{n}, (\mathbf{q} + \omega\mathbf{v})]^2}{(\mathbf{q}^2 - (q^0)^2 \epsilon_q)^2}\left(\frac{4\pi e_0}{q_1^2}\right)$$

$$\times \delta(q^0) Z_i^2 \left|\frac{1 - \tilde{\epsilon}_q^{l(e)}}{\tilde{\epsilon}_q^l}\right| \frac{\omega d\omega d\Omega_\mathbf{k} d\mathbf{q}}{(2\pi)^4}, \tag{6.15}$$

$$q^0 = \mathbf{q}\mathbf{v} + \omega - \mathbf{k}\mathbf{v}_0, \quad q_1^0 = q^0 - \omega, \quad \mathbf{q}_1 = \mathbf{q} - \mathbf{k}.$$

This cross section coincides with the polarization (transition) PBR cross section of fast particle on ion at rest in plasma for $\omega \gg \omega_{\text{pe}}$, determined by matrix element (21) from papers [73, 75].

Thus, the consistent quantum mechanical treatment of PBR in plasma has enabled to establish the unique physical nature of this phenomenon and studied earlier transition BR.

Now we consider the cross section that corresponds to the first term in curly brackets of Eqn. (6.13). It describes emission due to conversion of virtual photons of the IP field on individual plasma screened electrons. The term "individual" in this case means that the momentum – energy excess is transmitted to one plasma electron. The explicit form of the dielectric peremeability tensor components for the considered case $|q^0| \ll |\mathbf{q}|V_{Te}$, $|q^0| \gg |\mathbf{q}|V_{Ti}$:

$$\tilde{\epsilon}_q^l \approx 1 + \frac{1}{\mathbf{q}^2 r_{\text{D}}^2} - \frac{\omega_{\text{pi}}^2}{(q^0)^2}, \quad \tilde{\epsilon}_q^l(e) \approx 1 + \frac{1}{\mathbf{q}^2 r_{\text{D}}^2}, \quad r_{\text{D}} = \frac{V_{\text{Te}}}{\omega_{\text{Pe}}} \tag{6.16}$$

suggests the PBR cross section on individual plasma electrons be small at $|\mathbf{q}| < r_D^{-1}$. This corresponds to the well known in plasma physics fact of the long-wavelength Coulomb perturbations' screening [40].

If we rewrite the factor $\tilde{\epsilon}_q^{l(i)}/\tilde{\epsilon}_q^l$ entering the first term of Eqn. (6.13) in the form:

$$\frac{\tilde{\epsilon}_q^{l(i)}}{\tilde{\epsilon}_q^l} = 1 - \frac{\tilde{\epsilon}_q^{l(e)} - 1}{\tilde{\epsilon}_q^l} \qquad (6.17)$$

the first term in Eqn. (6.17) will describe the IP field scattering on the "naked" (not screened) electron, the second — on the polarization charge around it. In terms of papers [73, 75], the first type of radiation is the traditional BR on plasma electrons, the second — the polarization (transition) BR on them. The total emission is the superposition of these two channels. Interference between them drastically decreases the cross section at $|q^0| \ll |\mathbf{q}|V_{Te}$. This is explained by the polarization electronic charge around a singled-out electron having the opposite sign, and the IP field scattering amplitudes on the electron the polarization charge around it are close in the absolute value and opposite in sign if $|\mathbf{q}| \to 0$.

The situation pertinent to the IP field virtual photon scattering on ion in plasma is different. Here one can neglect scattering on the charge of the ion itself due to its large mass, so a non-compensated scattering on electronic polarization charge around ion remains. The above considerations make it clear why it is plasma ion (ion + its electronic "coating") that radiates at $|\mathbf{q}| < r_D^{-1}$, although at such momenta transmitted its effective charge is negligible: only charges with low mass (electrons of the Debye "coating") "show up" in radiation.

Thus the term in the total PBR cross section in plasma proportional to the electron dynamic form factor describes the superposition of polarization (transition) BR on plasma electrons and ions and static BR on plasma electrons.

Note that the contribution of bound electrons to the dielectric permeability for transition BR in rarefied plasma (the criterion will be given later) is unimportant. Indeed, the PBR cross section contains the factor $\left|\left(\tilde{\epsilon}_q^{l(e)} - 1\right)/\tilde{\epsilon}_q^l\right|^2 \approx (1 + |\mathbf{q}|^2 r_D^2)^{-2}$ and thus is small for $|\mathbf{q}| > r_{\mathrm{D}}^{-1}$. At the same time, energy – momentum conservation implies that $|\mathbf{q}| \geq \omega/\mathbf{V_0}$. These relations indicate that PBR on the Debye sphere is small within the frequency range $\omega > \mathrm{V_0}/r_{\mathrm{D}}$. Thus the spectral range where transition BR is significant in a low density plasma corresponds to low (in atomic scale) frequencies, for which the ionic core polarizability can be put equal to its static value. This is of the order of the cube of the ionic radius R_{i}. But then the addition to dielectric permeability from bound electrons proves to be insignificant, since the inequality $\omega_{\mathrm{pe}}^2/\omega^2 \gg n_{\mathrm{i}} R_{\mathrm{i}}^3$ holds at the considered frequencies. The situation changes for PBR on bound electrons belonging to the ionic cores, which will be considered below.

For a non-relativistic incident electron with simultaneous allowance for static and polarization channels in BR on the Debye sphere, the effect of "undressing" of plasma ion appears [76], which is similar to the atom "undressing" effect in the high frequency limit [77], when the under-logarithmic factor in the process's cross section coincides with that for a non-screened ion. The only difference is in the frequency intervals of the effect: for plasma $\omega_{\mathrm{pe}} \gg \omega < \mathrm{V_0}/r_{\mathrm{D}}$, for atom $I \gg \omega < \mathrm{V_0}/R_a$. If the incident particle is a positron, interference between static and polarization channels can lead to the differential (over the momentum transmitted) PBR cross section vanishing for $|\mathbf{q}| = r_{\mathrm{D}}^{-1}$, when $\epsilon_q^l = 2$, which also was predicted theoretically for PBR on atom [77].

Now we reproduce the result of calculations of ordinary (static) BR in plasma with transmission of the energy – momentum excess to a collective plasma excitation, a plasmon [78]. This process occurs near the zeros of the longitudinal part of the dielectric permeability in the expression for the photon propagator, as was noted above when discussing PBR.

The expression for the corresponding cross section can be derived using standard rules of quantum electrodynamics in a medium.

The following dispersion law for plasmon can be used [40]:

$$\kappa^0 = \omega_{\mathrm{pe}}\sqrt{1 + 3\frac{\kappa^2}{\kappa_{\mathrm{h}}^2}}, \quad |\kappa| \leq \kappa_{\mathrm{p}} = r_{\mathrm{D}}^{-1}. \quad (6.18)$$

In the non-relativistic limit for the IP motion, the differential cross section of BR accompanied with the generation of a plasmon can be found to be:

$$d\sigma_{\text{st}}^{\text{pl}}(k, q) = (2\pi)^{-3}\delta(q_1^0 + \omega_{\text{pe}}) \frac{d\omega}{\omega} \left(\frac{e_0^2}{m_0}\right)^2$$

$$\times \frac{\omega_{\text{pe}}}{V_0} \frac{[\mathbf{nq}]^2}{\mathbf{q}^2} (N_{-q}^{\text{pl}} + 1) d\Omega_{\mathbf{n}} d\mathbf{q}, \qquad (6.19)$$

$$|q| \leq r_{\text{D}}^{-1}.$$

Here N_{-q} is the occupation number of the plasmon mode with the 4-vector $-q$ which in the case of a non-degenerate plasma is $N^{\text{pl}} \approx T/\omega_{\text{pe}}$.

In this derivation, Eqn. (6.19) was integrated over the plasmon's wave vector with account of its dispersion (6.18). The spectral intensity of this process in non-degenerate plasma can be obtained from (6.19) to be

$$\frac{dW_{\text{st}}^{\text{pl}}(\omega)}{d\omega} = \frac{16}{3} \left(\frac{e^2}{m}\right)^2 \frac{V_{Te}}{V_0} r_{\text{D}}^{-3} \left(\frac{1}{9\pi}\right)$$

$$\times \left\{ 1 - \frac{1}{8} \left[1 + 3 \left(\frac{\omega}{\omega_{\text{pe}}}\right)^2 \left(\frac{V_{Te}}{V_0}\right)^2 \right] \right\}, \qquad (6.20)$$

$$\omega \leq \frac{V_0}{V_{Te}} \omega_{\text{pe}}.$$

In deriving (6.19) and (6.20) we have taken into account that the plasmon's wave vector is bounded from above by the value of the inverse Debye radius; at large Debye radii plasmons are not well-defined excited states. This results in the upper limit of the effect to appear $\omega^{\text{h}} = \frac{V_0}{V_{Te}} \omega_{\text{pe}}$. The numerical estimate of the ratio of spectral intensity (6.18) to that of static BR in plasma for the standard (according to Pippard) metal is approximately 5%. For a high frequency plasma mode in the alkali-haloid compounds, which is formed by valent electrons with higher plasma frequencies, the relative value of the effect under consideration must be larger.

Now we turn to consideration of bremsstrahlung radiation in plasma caused by the IP field conversion into a bremsstrahlung photon on bound electrons of the ion's core in plasma. In a low-density plasma at frequencies $\omega < V_0/r_D$ the most significant is emission on Debye "coatings" screening ions, while the ions themselves may be treated as point-like. However, for broader frequency range $\omega < m_e$ the contribution to BR from bound electrons should be taken into account.

The BR cross section due to bound electrons interaction with photons is given by the term of Eqn. (3.66) proportional to the dynamic form factor of the ion $S^{(i)}(q)$:

$$
d\sigma_{\text{i,bound}}^{\text{pol}}(\mathbf{k}, \mathbf{q}) = \frac{(q_1^0)^4}{V_0} \left| [\mathbf{nA}^{(0)}(q_1)] \right|^2 |c(\omega, \mathbf{k}, \mathbf{q_i})|^2
$$

$$
\times \left\{ |\delta n_i|_q^2 \left| \frac{\tilde{\epsilon}_q^{l(e)}}{\tilde{\epsilon}_q^l} \right|^2 + |\delta n_e|_q^2 \left| \frac{1 - \tilde{\epsilon}_q^{l(i)}}{\tilde{\epsilon}_q^l} \right|^2 \right\} \frac{\omega \, d\omega \, d\Omega_\mathbf{k} \, d\mathbf{q}}{(2\pi)^2}. \qquad (6.21)
$$

In deriving (6.21) spherical symmetry of the ionic core was assumed. Note that cross section (6.21) does not take into account the possible excitement of the ion's electronic subsystem during BR.

Plasma has a double effect on the process under study compared to the case of BR on an individual ion in vacuum. Plasma ions mutually screen each other, which is described by the factor before the first term in the curly brackets in (6.21). The second term in the curly brackets describes PBR with transmitting the energy – momentum excess from bound electrons of ions to plasma electrons due to interaction of ions with plasma electrons. Besides, interaction of the IP field with plasma produces a "dressing" of the corresponding virtual photon described by the photon propagator (3.47). This impact of the medium is included in Eqn. (6.21) as well. In a low-density plasma, the mutual screening of ions and the energy – momentum excess transmission from bound to plasma electrons can be neglected. Indeed, it is easy to show that these effects are important at $|\mathbf{q}| \approx |\mathbf{q_1}| \leq r_D^{-1}$, as follows from the form of ion's dynamic form factor (3.67) and expression for the longitudinal dielectric permeability in low frequency limit (6.16). But energy – momentum conservation demands such transmitted momenta be possible only at frequencies $\omega < V_0/r_D$ when polarizability of bound electrons is low. So for $|\mathbf{q}| < r_D^{-1}$ the corresponding bremsstrahlung radiation is weak. Polarizability of bound electrons takes a noticeable value (the core gets "defrozen") at $\omega > \omega_{\min}^{(i)}$ ($\omega_{\min}^{(i)}$ is the minimum exciting frequency of the core), but then $|\mathbf{q}| \approx |\mathbf{q_1}| > r_D^{-1}$, since for the typical parameters of non-degenerate plasma $\omega_{\min}^{(i)} \gg V_0/r_D$. So in Eqn. (6.21) one can assume

$$\left| \frac{\tilde{\epsilon}_q^{l(e)}}{\tilde{\epsilon}_q^l} \right|^2 = 1, \qquad \left| \frac{1 - \tilde{\epsilon}_q^{l(i)}}{\tilde{\epsilon}_q^l} \right|^2 = 0.$$

Bearing in mind the above considerations and the explicit form $\mathbf{A}^0(q_1)$, the PBR cross section on bound electrons in plasma can be found to have the form

$$
\begin{aligned}
d\sigma_{i,\text{bound}}^{\text{pol}}(\mathbf{k},\mathbf{q}) = {} & \frac{4\pi e_0}{V_0} |c(\mathbf{k},q_1)|^2 \delta(\omega + \mathbf{q}\mathbf{v_0} - \mathbf{k}\mathbf{v_0}) \\
& \times \frac{[\mathbf{n}(\omega\epsilon(\omega)\mathbf{v_0} - \mathbf{q})]^2}{(\omega^2\epsilon(\omega) - \mathbf{q_1}^2)^2} \frac{\omega^3 d\omega d\Omega_{\mathbf{k}} d\mathbf{q}}{(2\pi)^5}.
\end{aligned}
\qquad (6.22)
$$

From here for the spectral cross section at frequencies $\omega < V_0/R_i$ we obtain

$$
d\sigma_{i,\text{bound}}^{\text{pol}}(\omega) = \frac{16e_0^2}{3V_0^2} \frac{d\omega}{\omega} |\omega^2 \alpha_i(\omega)|^2 \ln\left[\frac{1}{\omega R_i (V_0^2 - \epsilon(\omega))^{1/2}} \right].
\qquad (6.23)
$$

In deriving Eqn. (6.23) scattering tensor was expressed through polarizability of the ionic core using Eqn. (3.20).

Compare cross section (6.23) with the transition BR on ions at rest in plasma at frequencies $\omega < \{V_0/r_D, \omega_{\min}^i\}$. The latter can be obtained from formula (6.15) after integrating over angular variables and the transmitted momentum module:

$$
d\sigma_i^{\text{pol}}(\omega) = \frac{16e_0^2 Z_i^2 e^4}{3m_e^2 V_0^2} \frac{d\omega}{\omega} \ln\left[\frac{\gamma V_0}{r_D \sqrt{\omega^2 + \gamma^2 \omega_{pe}^2}} \right].
\qquad (6.24)
$$

At frequencies considered, the ion's polarizability can be approximated by its low frequency limiting value: $\alpha_i(\omega \to 0) \to R_i^3 \approx (me^2)^{-3} Z^{-1}$. Then, neglecting the difference of logarithmic factors in cross sections, we obtain the desired ratio:

$$
\zeta(\omega) = \frac{d\sigma_{i,\text{bound}}^{\text{pol}}}{d\sigma_i^{\text{pol}}} \approx \frac{\omega^4}{(me^4)^4 (ZZ_i)^2}, \qquad \omega < V_0/r_D, \omega_{\min}^{(i)}.
\qquad (6.25)
$$

It follows from here that PBR on bound electrons is small with respect to emission on free electrons for ($\zeta(\omega) \ll 1$) frequencies

$$\omega \ll \sqrt{ZZ_i}\, me^4 \approx \omega_{\min}^{(i)}. \tag{6.26}$$

So, if the inequality holds

$$V_0 \ll r_D \omega_{\min}^{(i)} \tag{6.27}$$

PBR on bound and plasma electrons are essential within different (non-overlapping) frequency intervals and interference between them can be ignored.

Let us assess now at which plasma number densities (n^*) this interference becomes significant. For this we bear in mind that PBR on the ion's core is important at $\omega > \omega_{\min}^{(i)} \approx me^4 \sqrt{ZZ_i}$. Radiation on plasma electrons has an upper frequency of the order of $V_0 \sqrt{4\pi n e^2/T_e}$. As a result, at electron temperatures of the order of the ionization potential of the ion, the number density sought for is:

$$n^* \approx 10^{19} ZZ_i^3 \,\mathrm{cm}^{-3} \tag{6.28}$$

Then the criterion of the "low-density" of plasma, when interference between PBR on plasma and bound electrons can be disregarded, is given by the inequality $n < n^*$, where the number density n^* is determined by Eqn. (6.28).

Below we present the expression for the PBR spectral cross section on the ionic core in plasma within the frequency interval $V_0/R_i \gg \omega \gg I$ (I is the maximum ionization potential of the ion), when one can make use of the high frequency limit for dynamic polarizability of bound electrons:

$$d\sigma_{i,bound}^{pol} = \frac{16e_0^2 e^4 (Z - Z_i)^2}{3m^2} \frac{d\omega}{\omega} \ln\left[\frac{\gamma}{R_i\sqrt{\omega^2 + \gamma^2 \omega_p^2}}\right],$$

$$\omega_p = \sqrt{\frac{4\pi n_i Z e^2}{m_e}}. \tag{6.29}$$

It is interesting to compare Eqn. (6.29) with formula for PBR on the Debye "coating" of plasma ion (6.24). These expressions coincide if make substitutions: $Z - Z_i = N_{be}$, $R_i \to r_D$, $\omega_{pe} \to \omega_p$.

It is clear from here that PBR on bound electrons at $\omega \gg I$ is fully analogical to PBR on the Debye "coating" of plasma ions; only in this case the electronic "cloud" around ionic core plays the role of the polarization "coating". Such a correspondence is explained by bound electrons interacting with an electromagnetic field at $\omega \gg I$ as free particles, since their eigenfrequencies are much smaller than the field frequency. The coupling with the core then appears only in the spatial localization of bound electrons. If $\omega \ll I$, the behavior of bound electrons is determined by quantum mechanical laws, in particular their resonance excitation becomes possible, etc.

From the viewpoint of the correspondence of terminology accepted in the theory of PBR on atom to the case of transition BR, we can note that that the latter process on a plasma ion is "elastic" PBR on the Debye sphere, while the Compton BR on plasma electrons (see [73, 75]) corresponds to "inelastic" PBR on the Debye sphere.

Equation (6.23) differs from its analog in the case of PBR on single atom (ion) only by under-logarithmic expression. This, in particular, ensures that PBR on an ionic core in a low-density plasma not be suppressed by the density effect, as well as the transition BR [73, 75]. Both these facts admit one and the same explanation: a photon is emitted by non-relativistic particles — plasma and bound electrons. So the increase in the electromagnetic wave phase velocity in plasma (which entails the density effect for ordinary BR) does not affect the probability of the process. (The density effect for PBR discovered in paper [79] relates to the case of dense plasma, when the ion number density is of the order of the reverse ion's volume; this effect is due to the destructive interference of contributions from different atoms of the medium).

The change in the under-logarithmic expression in Eqn. (6.23) compared to PBR in vacuum is caused by the "dressing" of the IP field in the medium. Here we should note that for plasma ions with a core within the frequency interval $\omega_{min}^{(i)} < \omega < I$, dielectric permeability of plasma can be larger than unity. Then the phase velocity of a photon in the medium decreases and an effect inverse to the density effect for ordinary BR should take place. For polarization BR this "dressing" results in the IP field strengthening instead of decreasing. If the denominator in the under-logarithmic expression of (6.23) vanishes, the Vavilov – Cherenkov condition is fulfilled. PBR in such a situation has been discussed by some authors [79, 80].

Thus PBR on the ionic core in a "low-density" plasma virtually coincides with the process in vacuum and its interference with transition BR can be neglected.

The last statement remains valid for relativistic IP as well, when at small emission photon angles a compensation of the excess momentum transmitted to the target during BR due to the emission photon momentum becomes possible. In this case the frequency ranges, where PBR on the Debye sphere and bound ion's electrons is important, can overlap in a low density plasma (at sufficiently high IP energies), too, provided that $\omega_{\min}^{(i)} < \gamma^2 r_D^{-1}$. However, interference between these two channels remains negligible since it is "smeared out" by the contribution of large enough transmitted momenta: $r_D^{-1} < |\mathbf{q}| < R_i^{-1}$, when PBR on free electrons is small and on bound electrons is large.

Figure 21(a) displays the PBR intensity spectral dependencies for a zero photon emission angle on an ion structure in a plasma with the Debye radius $r_D = 10^5$ a.u. for two values of the IP energy. It is evident that on the curve corresponding to the smaller energy ($\gamma = 1.05$) a dip appears, because the frequency ranges where PBR on free and bound electrons is significant do not overlap. In contrast, at higher energy ($\gamma = 5$) the inequality $\omega_{\min}^{(i)} < \gamma^2 r_D^{-1}$ holds, and these ranges overlap so that no dips appear on the curve of the PBR spectral intensity. In this case interference between PBR on the ionic core and Debye sphere is negligible due to a large difference between the Debye radius and the ionic core size.

In Figure 21(b) we show the (normalized) PBR spectrum in a dense solar plasma with parameters $r_D = 1$ a.u., $Z_i = 15$, $T = 57$ a.u. for a zero photon emission angle. The static polarizability of bound electrons of plasma ions is assumed to be $\alpha_i(\omega \approx 0) = 10^{-3}$ a.u., and the relativistic IP factor is taken to be $\gamma = 10$.

For these parameters the spectral ranges inside which PBR on free and bound electrons is important overlap.

The solid curve in Fig. 21(b) includes the ionic core contribution to the process cross section; the dashed line corresponds to the point-like ion approximation. Clearly, in the latter case destructive

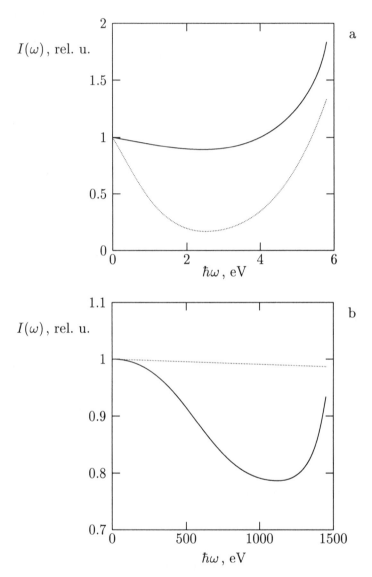

Fig. 21. Normalized spectral PBR intensities of a fast particle in plasma on an ion with $\alpha(0) = 1$ a.u. for zero emission angle (a) $Z_i = 1$, $r_D = 10^5$ a.u., $\gamma = 5$ (the solid curve) and $\gamma = 1.05$ (the dashed curve). (b) The same for a dense (solar) plasma: $Z_i = 15$, $r_D = 1$ a.u., $T = 1.55$ keV, $\gamma = 10$.

interference between two PBR channels, corresponding to the gap on the solid curve, leads to the 20% decrease in the total cross section. The increase of the interference contribution is due to the difference in size between the Debye "coating" and the ion's core being approximately one and a half order of magnitude smaller in this example over the previous one (Fig. 21(a)).

Note that the solar plasma considered still stays ideal and non-degenerate.

Thus we can conclude that interference between PBR on free and bound electrons in the ideal non-degenerate plasma can be significant only for large densities.

6.3. Interference – polarization effects during radiation of relativistic particles in a dense medium

It is well known (see, for example, [81]) that during the ordinary (static) BR of relativistic particles in a medium interference effects arise due to the quadratic increase of the radiation formation length with IP energy $L \approx \gamma^2 \lambda$ (γ is the relativistic Lorentz factor, λ is the wavelength). When the radiation formation length exceeds the mean interparticle distance, processes of scattering of both the electron itself (the Landau – Pomeranchuk effect) and the photon (the density effect) on the medium particles begin to affect the probability of the elementary emission act. In addition, within crystals collective effects are possible due to the emitting particle entering the channeling regime and transferring the momentum excess to the lattice as a whole (coherent BR). There are also some interference effects caused by the sample's boundaries.

In the case of the polarization channel, emission of a photon results from scattering of the IP proper field into a real photon on non-relativistic bound electrons. So the medium has, as a rule, a smaller effect on the polarization mechanism of relativistic particles. This fact is clearly demonstrated in BR in plasma. Then the static channel turns out to be suppressed in the low frequency range $\omega < \gamma \omega_{\mathrm{pe}}$ (ω_{pe} is the electron plasma density) because of the process's formation length decrease due to the increase of the phase velocity of light in plasma — the Ter-Mikaelian effect [81]. At the same time, the increase of the phase velocity of electromagnetic radiation has virtually no effect on the polarization BR cross section which in partially ionized plasma can occur both on the Debye sphere (transition BR) [73, 75] and on the bound electrons of ions [76]. The effect of the medium here is reduced to screening the IP proper field, which changes the under-logarithmic factor in the cross section. As a result, logarithmic increase of the polarization BR cross section with the IP energy, which would take place for the process on an isolated atom, stops at frequencies $\omega < \gamma \omega_{pe}$. Therefore, the medium impact on the polarization channel of BR of relativistic particles in plasma proves to be much more weaker than for ordinary BR.

The situation, however, changes for polarization BR of relativistic particles in an amorphous condensed matter. Then, as was first shown in [79], the polarization channel is strongly suppressed by destructive interference of contributions from chaotically located atoms into the cross section of the process. By its nature, this effect is similar to the well known effect of X-ray scattering cross section decrease in a condensed medium in the low frequency range [79]. Indeed, as we already discussed, the polarization BR can be interpreted, in the spirit of the equivalent Fermi photons, as being the IP proper field scattering into a real photon on target's electrons, so the characteristic features of this scattering will be reflected in the PBR differential cross section (the effect of the medium). After averaging over contributions from different atoms, the following additional factor appears in the cross section [79]:

$$M(\mathbf{q}_1) = 1 - \sigma \, \frac{3 \, j_1(x)}{x}, \quad \sigma = \frac{4 \, \pi \, n_0 \, R_0^3}{3}, \quad x = |\mathbf{q}_1| \, R_0, \qquad (6.30)$$

where $j_1(x)$ is the spherical Bessel function, $\mathbf{q}_1$ the momentum transferred to the medium, n_0 is the medium's atomic number density, and R_0 is the average atomic size. Equation (6.30) indicates that the destructive interference considered is essential only for a sufficiently dense medium when the parameter σ is of order unity. Besides, the inequality $x < 1$ should be met, thus imposing a certain restriction on the bremsstrahlung photon frequency. An analysis carried out in [79] showed that the maximum frequency ω_h, which restricts from above the appearance of the density effect in PBR, also depends on the ratio of the screening radius (the Thomas – Fermi radius in the statistical model) to the atomic size: for a zero emission angle, the maximum frequency ω_h increases with this ratio decreasing, and in the opposite case $\omega_h \to c/R_0$.

In a periodic medium, in addition to binary collisions the IP momentum excess during photon emission can also be transmitted to the lattice as a whole (coherent PBR). Here the role of this process for the polarization channel proves to be even more significant than for the ordinary (static) channel [82]. The point is that coherent BR is produced for sufficiently small momenta transmitted, less than the inverse thermal fluctuation of the atomic locations u_T: $|\mathbf{q}_1| < u_T^{-1}$ [79]. Polarization mechanism is also significant for small q_1: $|\mathbf{q}_1| < R_0^{-1}$. Since $u_T \ll R_0$, the necessary condition for polarization BR is excessively fulfilled. One can say that at frequencies $\omega < c/u_T$ polarization BR occurs mainly with a transfer of momentum to the lattice as a whole, and only for sufficiently high frequencies $\omega > c/u_T$ non-coherent PBR dominates. The same effect takes place for PBR in polycrystals as well [82].

7. Polarization – interference phenomena in radiation of thermal electrons in a low-temperature plasma

Now we turn to considering polarization and interference effects in bremsstrahlung emission (absorption) of photons in scattering of slow electrons with velocities

$$v_0 \ll v_a \qquad (7.1)$$

on atoms (v_a is the characteristic tic velocity of atomic electrons). In this case the plane wave approximation is, strictly speaking, inadequate to describe the IP motion, so other methods of calculations are required. The bremsstrahlung effect in scattering of slow electrons on neutral atoms was first considered in papers [83] and [84]. In paper [83], bremsstrahlung emission cross section and absorption for electrons with energies $E < 3$ eV were expressed through the elastic scattering cross section. Bremsstrahlung absorption of infrared radiation in electron scattering on the hydrogen atom was calculated in paper [84]; there the corresponding matrix element of the transition was expressed through the phase shift of the s-wave of the IP wave function.

The need for taking into account a neutral atom's polarization by scattering slow electron was noted in [83], where the ratio of the static to polarization emitting dipole moments was estimated using a classical model for electron motion in atom's field. This estimate suggested that the radiation occurs via polarization channel at least at the IP distances from the nucleus of the order of the atomic size. Nevertheless, in calculating the radiation intensity in gas and the absorption coefficient only the static channel was accounted for.

The polarization term in the bremsstrahlung emission amplitude of a slow electron on atom was obtained within quantum mechanical approach in [63] using a diagram technique. For a slow electron, condition (7.1) and energy conservation lead to the obvious inequality for the emitted photon frequency $\omega \ll \omega_a$, where ω_a has the sense of the minimum eigenfrequency of atomic electrons. Considering this fact the following estimate was obtained for the PBR cross section within the frames of the first Born approximation

$$\frac{d\sigma^{(\mathrm{pol})}}{d\omega} \approx \frac{8\,\omega^3}{9\,c^3\,E_0}\,|\beta_{il}|^2 \ln\left(\frac{k_0 + k}{k_0 - k}\right), \qquad (7.2)$$

where E_0 is the initial energy of the scattering electron, β_{il} is the atom's static polarizability tensor, k_0, k the initial and final momentum of the incident electron, respectively. The static PBR cross section in this approximation can be presented in the form

$$\frac{d\sigma^{(\mathrm{st})}}{d\omega} = \frac{256\,\pi^2\,E_0}{27\,\omega\,c^3}\,\left|\langle a_0|r^2|a_0\rangle\right|^2. \qquad (7.3)$$

Here $|a_0\rangle$ is the wave function of atom's core. Note that the use of the first Born approximation to assess the the BR cross section of slow electrons on neural atoms proves to be more justified (see below) for the polarization channel than for the static one.

PBR spectral cross section (7.2) was shown in [63] to attain a maximum value at frequency $\omega_{\mathrm{opt}} \approx 0.8\,E_0$ with vanishing at small frequencies and at the boundary frequency $\omega = E_0$. Comparison of contributions from both channels (7.2) and (7.3) at the frequency $\omega = \omega_{\text{opt}}$ for $\beta_{il} \approx \langle r^2 \rangle / I_a$ (I_a is atom's ionization potential) indicates that the static radiation intensity is by two orders of magnitude larger than the corresponding PBR intensity. Although basic formulas (7.2) and (7.3) are approximate, using possible more accurate expressions does not principally alter the results.

Thus we can conclude that core polarization effects in bremsstrahlung of slow electrons (in the sense of inequality (7.1)) on neutral atoms are small compared to the classical estimation carried out in [83]. The reason for this is in the emission frequencies of slow electrons being small. At such frequencies, the atom's core stays "not defrozen" and the polarization radiation intensity is low, as follows from Eqn. (7.2) where the PBR spectral cross section is proportional to the cube of the frequency. Note that in approaching the bremsstrahlung photon energy to an atom's ionization potential the dispersion of polarizability should be taken into account, so evaluation using Eqn. (7.2) proves inapplicable in this frequency range.

Unlike emission, the above conclusion does not relate to the bremsstrahlung absorption of photons by slow electrons, whence the energy of the absorbed photon can be significantly larger than the initial energy of the scattering electron. The corresponding numerical calculations with a correct accounting for both radiative channels were first carried out in [85].

The basic expression for the BR amplitude including polarization effects, which was used in [85], has the form (see also [9, Chap. 7]):

$$F_{\varepsilon,\varepsilon+\omega} = \langle \varepsilon | (\mathbf{e}\,\mathbf{d}) | \varepsilon + \omega \rangle$$

$$+ 2 \sum_{\substack{j \leq F \\ \varepsilon'' > F}} \langle \varepsilon\, j | u | \varepsilon + \omega,\, \varepsilon'' \rangle \frac{(\varepsilon'' + I_j)\, \langle \varepsilon'' | (\mathbf{e}\,\mathbf{D}(\omega)) | j \rangle}{\omega^2 - (\varepsilon'' + I_j)^2}\,. \quad (7.4)$$

Here u stands for the operator of the IP interaction with the electronic core of the target and other notations are the same as in Eqn. (2.32). The first term in Eqn. (7.4) represents the static channel amplitude for the inverse bremsstrahlung effect (bremsstrahlung absorption); the second term gives the polarization channel contribution to absorption. The effective dipole moment $D(\omega)$ can be found, as mentioned earlier in Sect. 2, in the random phase with exchange approximation, which takes into account effects of inter-electron interactions in target's core during the radiative process.

If, as is the case of the situation considered, the energy of the absorbed photon notably exceeds the initial electron energy (it is in this case that the relative contribution of the polarization term can be significant), then for a qualitative analysis the total process amplitude (7.4) can be simplified to be

$$F_{\varepsilon,\varepsilon+\omega} = \frac{4\pi p\, p'\,(\mathbf{e}\,\mathbf{p}')}{\omega} \left[\frac{a}{\omega} + \alpha_{\mathrm{d}}(\omega) \right], \qquad (7.5)$$

where a is the electron scattering length on atom. Note that in deriving Eqn. (7.5) we have also assumed that the electron momentum in the final state is smaller than the typical momentum of atomic electrons. The first term in Eqn. (7.5) describes the static channel and depends upon the ability of atom to scatter charged particles. The second term is proportional to dynamic polarizability of the atom and accounts for the polarization channel. It is essential that both terms similarly depend on the transmitted momentum, unlike the case of fast particles (see Sect. 6), where the static amplitude increases with the momentum transmitted and the polarization amplitude decreases. The reason for such a behavior is that a slow IP has insufficient energy to effectively penetrate into the electronic core of an atom, so the polarization term is determined by dipole polarizability while the static term is established by the scattering phase of the s-component of the electron's wave function.

This peculiarity of the amplitude dependence of the channels on the momentum transmitted makes it possible for the cross-channel interference to appear not only in the differential over the electron scattering angle cross section (as is the case for BR of fast IP), but also in the integral over the angle bremsstrahlung effect cross section which depends only on the frequency. For example, for negative values of the scattering length corresponding to effective attraction of electron to atom, the total bremsstrahlung absorption amplitude (7.5) can vanish for frequencies $\omega < I_p$, at which dynamic polarizability is positive. The estimation of this dip frequency can be obtained from Eqn. (7.5) to be

$$\omega_{\min} \approx |a|/\alpha_d(0) . \tag{7.6}$$

For example, for argon $\omega_{\min} = 0.26$ Ry, for xenon $\omega_{\min} = 0.4$ Ry. As in both cases the interference dip frequency proves to be less than the first exciting potential of the atom – target, estimate (7.6) can be considered to be sufficiently correct.

For positive values of the scattering length that correspond to effective repulsion of IP from target, total bremsstrahlung absorption amplitude (7.5) at frequencies below minimum eigenfrequency of target's core does not vanish since both terms have the same sign.

At frequencies above an atom's ionization potential, the dynamic polarizability, and the total bremsstrahlung effect amplitude as well, become complex quantities. No interference-induced vanishing of the corresponding cross sections occurs and only a shallow minimum may appear in the spectral dependence of the bremsstrahlung absorption coefficient.

As seen from Eqn. (7.5), relative contribution of the polarization channel to the total bremsstrahlung absorption of a photon by a slow electron on neutral atom (R-factor) reads

$$R^{(\text{slow})} = \left(\frac{\omega \, \alpha_d(\omega)}{a}\right)^2 . \tag{7.7}$$

Equation (7.7) entails that the relative contribution of the polarization channel in the bremsstrahlung absorption cross section in slow electron scattering increases quadratically with the photon frequency. Thus it is natural that at low frequencies corresponding to the low electron BR this contribution is small, which is consistent, clearly, with estimates from paper [63].

Substituting in Eqn. (7.7) the static atom's polarizability $\alpha_d(0) \approx N_{\text{ext}}/I_p^2$ (N_{ext} is the number of external electrons) and assuming the scattering length in the form $|a| \approx \langle r \rangle \approx I_p^{-1}$, the polarization-to-static cross section ratio takes the form

$$R^{(\text{slow})} \approx (N_{\text{ext}} \, \omega/I_p)^2 . \tag{7.8}$$

Obviously, the polarization channel contribution at frequencies $\omega \approx I_p$ (where an order of magnitude estimation is valid) is large and for $N_{ext} > 1$ can become dominating. It should be born in mind that at frequencies $\omega \approx I_p$ the condition that the final electron momentum module is small compared to the characteristic atomic momentum is violated. This has been used to obtain the basic formula (7.5) from original expression (7.4). So for a correct understanding of the role of the polarization channel in this case one should carry out an accurate numerical calculation of the photon absorption cross section using general formula (7.4). Such a calculation was carried out in paper [85] for slow electron scattering on the argon and xenon atoms in external radiation field. The process cross section integrated over directions of the final momentum of accelerated electron was presented as a function of the external field frequency in the spectral range 0.5 – 1.5 Ry for two initial electron energies $\varepsilon = 0.01, 0.09$ Ry with (and without) allowance for polarization channel. Polarization effects in the bremsstrahlung absorption spectral cross section were found to be large not only in the near-resonance range but practically within the entire frequency range considered, with total-to-static channel ratio for the argon atom being around 3. The cross section for a higher-energy electron ($\varepsilon = 0.09$ Ry) proved to be about 10% larger over a lower-energy electron ($\varepsilon = 0.01$ Ry).

Note that in these calculations of the bremsstrahlung absorption cross section only partial s- and p-waves ($l = 0, 1$) were included, which was enough in view of small IP energies. For frequencies of the order of atom's ionization potential $\omega \approx I_p$ the amplitudes of the channels were found to have opposite signs, in agreement with estimate (7.5).

An analytical description of the inverse bremsstrahlung effect for slow electrons on neutral atoms with an account for polarization channel was first obtained in paper [59]. In particular, it was shown that the plane wave approximation for the IP wave function is appropriate to calculate the polarization channel contribution to the process cross section even in the case of slow electrons. In these calculations the initial p and final p' momenta of the accelerated electron were assumed to be small compared to the atomic momentum p_a. The total bremsstrahlung absorption amplitude (in atomic units) is obtained to be

$$A_{\mathbf{pp'}}(\omega) = 4\pi^2 i\,(\mathbf{M}\,\mathbf{E}_0)\,\delta(\varepsilon + \omega - \varepsilon')\,, \qquad (7.9)$$

where

$$\mathbf{M} = \frac{1}{2\omega^2}\left(\frac{\mathbf{p'}}{p}\sin(\delta_0) - \frac{\mathbf{p}}{p'}\sin(\delta_0')\right) + \frac{\mathbf{q}}{q^2}\beta(\omega)\,. \qquad (7.10)$$

Here $\mathbf{q} = \mathbf{p} - \mathbf{p'}$ is the change in the scattering electron momentum, $\beta(\omega)$ is the atom's dipole polarizability, and δ_0 is the momentum-depending scattering phase of the s-component of the electron wave function. Other partial waves were ignored due to the scattering electron energy smallness.

Note that if the scattering phases of the accelerated electron is set to zero, the first term in Eqn. (7.10), which describes the static channel, vanishes. This corresponds, clearly, to a free charge in a vacuum being incapable of absorbing a photon. At the same time, the last term in Eqn. (7.10) stays non-zero for a free IP as well, as was discussed above.

Formulas (7.9), (7.10) yield the simple expression for the absorption spectral cross section depending only on the initial electron velocity [59]:

$$\sigma_\omega(v) = \frac{16}{3}\frac{\pi^2}{137}\frac{v'}{v}\left\{\frac{1}{2\omega^3}\left(\frac{v'^2}{v^2}\sin^2(\delta_0) + \frac{v^2}{v'^2}\sin^2(\delta_0')\right)\right.$$

$$\left. + \frac{\beta^2\,\omega}{v\,v'}\ln\left(\frac{v'+v}{v'-v}\right) + T_{\text{inter}}\right\}\,, \qquad (7.11)$$

where the first term in the curly brackets corresponds to the static channel, the second to the polarization channel, and the third represents the interference term (which has an awkward explicit form so we omit it here). Setting $\beta = 0$ in Eqn. (7.11) yields the result [83] that was found ignoring the polarization channel.

A numerical estimation of the radiation absorption spectral coefficient by ionized gas, determined by the standard expression $k_\omega = N_e N_a\langle v\,\sigma_\omega(v)\rangle$, can be obtained from Eqn. (7.11) after averaging over the Maxwell electron velocity distribution. The scattering phase of the s-wave, entering Eqn. (7.11), was computed using the O'Melly – Sprach – Rosenberg formula [86]. Radiation absorption in neon and argon was considered as a function of frequency for different gas temperatures. Static atom polarizabilities were used $\beta_{Ne} = 0.4$ Å^3, $\beta_{Ar} = 1.64$ Å^3. For comparison with relationships obtained in paper [59], the curves for spectral absorption coefficients were calculated and plotted ignoring atom – target's polarization. The contribution of polarization effects in radiation absorption was found to be small for neon, explained by its low polarizability. In contrast, polarization and interference effects for the absorption coefficient in argon are large. In particular, these effects cause a broad minimum to appear on the absorption spectral curve (instead of a monotonic decrease typical for the low-polarizability atoms). The depth of this interference dip increases with decreasing gas temperature and somewhat shifts toward high frequencies. For a temperature of $T = 300°$ K, the interference minimum in argon locates at about $\omega_{min} = 5 \cdot 10^{15}$ Hz, i.e. in the optical range.

The dip in the spectral absorption coefficient in argon, caused by the destructive cross-channel interference in the bremsstrahlung absorption by slow electrons discussed above (see Eqn. (7.5)), is also due to the negative sign of the electron scattering length on the argon atom ($a_{Ar} = -0.875$ Å). Note that the scattering length of slow electron on the neon atom is positive ($a_{Ne} = 0.106$ Å). The interference dip frequency can be found in correspondence with the quantity determined by formula (7.6) and it significantly exceeds Ramsauer's dip width in the elastic electron scattering on atoms. So the minimum calculated on the photoabsorption curve does not directly relate to the Ramsauer effect and is a consequence of the destructive interference of static and polarization channels.

Polarization mechanism in radiative processes in scattering of slow electrons on ions with an electronic core has been first considered in monograph [9]. The slowness of electrons in this case, determined by condition (7.1), coincides with the condition of quasi-classical motion in the Coulomb field. The corresponding inequality is opposite to the Born approximation. Radiative processes in scattering quasi-classical electrons for sufficiently high frequencies of the emitted (absorbed) photon (see below frequency condition (7.13)) are effectively described by methods of the so-called Kramers electrodynamics [47]. The physical picture underlying this method is that during emission of high-frequency photons by a quasi-classical electron in the atomic potential $U(r)$, the radiation is formed in a narrow region of the radial electron motion near the classical turn-off point. The corresponding radiation radius (r_ω) is a solution to the equation

$$\frac{v_{\mathrm{i}}^2}{2} + |U(r)| = \frac{\omega^2 r^2}{2} \tag{7.12}$$

and the emitting frequencies are close to the electron's rotation frequency on the trajectory near this point:

$$\omega \approx \omega_{\mathrm{rot}}(r_\omega) = \left[\left(v^2 + 2|U(r_\omega)|\right)\Big/ \mathrm{r}_\omega^2\right]^{1/2}. \tag{7.13}$$

For the Coulomb field Eqn. (7.12) can be solved explicitly to give $r_\omega \approx (Z/\omega^2)^{1/3}$. At such small distances r_ω, the electron strongly accelerates so that its trajectory and (due to localization of the radiating part of the trajectory at the turn-off point) the emission spectra do not depend on the initial electron velocity and are determined by the angular momentum only. The independence on the initial energy allows the formula for the radiation intensity to be applied for electron transitions both in the continuum and to the discrete spectrum, i.e. for recombination, as well as for transitions in the discrete spectrum as well [47].

To calculate the PBR spectral cross section, one may apply Fermi's method of equivalent photons [22], according to which the corresponding intensity is represented as a product of the radiation scattering cross section on the ion's electronic core by the flux density of equivalent photons linked to the scattering electron's field.

Formula for the equivalent photon flux density is determined by the well-known Fourier components of the electron's trajectory in the Coulomb field and for strongly distorted trajectories this is expressed in terms of the angular momentum of the slow electron (M) in the form [9, Chapt. 11]:

$$I_\omega = \frac{137\,\omega\,M^4}{6\,\pi^2\,Z^4}\left\{K_{1/3}^2\left(\frac{\omega\,M^3}{3\,Z^2}\right) + K_{2/3}^2\left(\frac{\omega\,M^3}{3\,Z^2}\right)\right\}, \tag{7.14}$$

where $K_{1/3}$ and $K_{2/3}$ are the McDonald functions.

Note that the possibility of describing by the classic formula both the BR and photorecombination process comes from the true condition of the spectrum quasi-classicality being local due to the spatial localization of the radiating region and thus depending upon the local kinetic energy $\varepsilon_{\text{kin}}(r_\omega)$ of the accelerated electron at quantum's emission point:

$$\omega/\varepsilon_{\text{kin}}(r_\omega) \ll 1. \tag{7.15}$$

The differential (over the impact parameter ρ or the angular momentum M) PBR intensity $Q_\omega^{(\text{pol})}$ is determined, as noted above, by the product of the Fermi equivalent photon flux

$$Q_\omega^{(\text{pol})} = I_\omega \, \sigma_{\text{scat}}(\omega) \tag{7.16}$$

by the electromagnetic field scattering cross section on the core of ion – target $\sigma_{\text{scat}}(\omega)$. From [87], we obtain

$$\sigma_{\text{scat}}(\omega) = \frac{8\,\pi\,\omega^4}{3\,c^4}\,|\alpha(\omega)|^2. \tag{7.17}$$

Thus Eqns (7.14), (7.16), and (7.17) in fact give the solution to the problem of the slow electron PBR on ions in high frequency range (7.13). Moreover, Eqns (7.16) and (7.17) also describe PBR in the low frequency range (in the case opposite to (7.13)), provided that the equivalent photon flux density is taken in the form corresponding to a weak deviation of the initial straight electron's trajectory

$$I_\omega = \frac{137\,\omega}{2\,\pi^2\,v^4}\left\{K_0^2\left(\frac{\omega\,\rho}{v}\right) + K_1^2\left(\frac{\omega\,\rho}{v}\right)\right\} \tag{7.18}$$

which was used by Fermi [22] in calculating atomic excitation cross sections by a charged particle with impact parameter ρ.

In order to obtain the total spectral cross section of the polarization BR, formula (7.16) should be integrated over the angular momentum of the scattering electron or, equivalently, over the impact parameter.

It is important to emphasize that the PBR cross section for slow electrons on ions expressed through ion's dipole polarizability (7.16), (7.17) is justified when the radius responsible for the equivalent photon emission with frequency ω exceeds the ionic core size r_i.

For the Coulomb field the condition for the dipole PBR is given by the inequality $\omega < (Z/r_i^3)^{1/2}$. The consistency of the latter with high-frequency condition (7.13) in the Coulomb field yields the following bound for the IP energy: $\varepsilon < Z/r_i$, which is exactly the criterion of an electron's "slowness". The conditions for the scattering electron dipole interaction with target's core being met, the R-factor characterizing the relative contribution of the polarization channel reads

$$R = \left[\frac{\omega^2 \, \alpha(\omega)}{Z} \right]^2, \qquad (7.19)$$

which is well known in the polarization bremsstrahlung radiation theory [9].

8. Polarization radiation and absorption in a laser field

8.1. Multiphoton polarization bremsstrahlung emission and absorption

Multiphoton processes, as a rule, have an appreciable probability in a sufficiently intensive external field. This problem became especially actual in connection with laser technique development which can generate very powerful electromagnetic pulses with an electric field strength of the same order as or even exceeding the atomic.

Multiphoton BR on a Coulomb center (without polarization channel) for fast electrons was first calculated in [89]. In the Born approximation for the incident electron, the following familiar expression for the process cross section was obtained:

$$d\sigma_n^{\text{st, Born}}(\Omega) = J_n^2(\mathbf{aq}) \, d\sigma^{\text{Coul}}(\Omega). \qquad (8.1)$$

Here J_n is the Bessel function of the n-th order; $\mathbf{a} = \mathbf{E}_0/\omega^2$ is the amplitude of electron oscillations in the laser field with frequency ω and amplitude $\mathbf{E}_0$, $\mathbf{q}$ is the change of the electron's momentum in the inelastic scattering into the solid angle Ω, and $d\sigma^{\text{Coul}}(\Omega)$ is the Coulomb scattering cross section. In the opposite case of quasi-classical electron motion, in [90] an expression with similar structure was derived, in which $\mathbf{q}$ is substituted by the product $\omega \mathbf{v}_\omega$, where $\mathbf{v}_\omega$ is the Fourier component of the classical IP trajectory in the Coulomb center's field.

Multiphoton bremsstrahlung effect of fast electrons on atoms in a strong electromagnetic field with account of the polarization channel was first considered in paper [91] within the frames of the first Born approximation. The corresponding differential cross section for the s-photon process has the form (in atomic units)

$$
\begin{aligned}
d\sigma^{(s)}(\mathbf{p}, \mathbf{p}') = \frac{4}{q^4} \frac{p'}{p} \Big| & J_s(\rho) \left[Z - F(\mathbf{q}) \right] \\
& - \frac{1}{2} \rho \, \omega^2 \, \alpha(\omega) \left[J_{s-1}(\rho) - J_{s+1}(\rho) \right] \Big|^2 d\Omega_{\mathbf{p}'} .
\end{aligned} \tag{8.2}
$$

Here $\rho = \mathbf{q} \, \mathbf{E}_0 / \omega^2$, $F(\mathbf{q})$ is the atomic form factor. The first term inside the modulus describes the contribution from the static channel; it coincides with Eqn. (8.1). The second term is the contribution due to the target's polarization, i.e. related to the polarization channel.

As was noted in [91], for $|s| \geq 2$ Eqn. (8.2) becomes, generally speaking, inaccurate, since it does not take into account higher-order (above the first one) interactions of atomic electrons with the external electromagnetic field. So result (8.2) describes emission (absorption) by the polarization channel of only one photon, other photons being "supplied" through the static channel.

Higher orders of atomic interactions with external electromagnetic field lead to the appearance of the cross section dependence on the non-linear atomic polarizability. They have been taken into account in [92], where the differential cross section of the induced bremsstrahlung effect with account of the target's polarization has been obtained in the form

$$
\frac{d\sigma^{(s)}}{d\Omega_{\mathbf{p}'}} = \frac{4 \, p'}{p \, q^4} \Big| \sum_{n+m=s} \left[-Z \, \delta_{n0} + F_n(\mathbf{q}) \right] J_m(\rho) \Big|^2 . \tag{8.3}
$$

Here Z is the nuclear charge; $F_n(\mathbf{q}) = \int n_n(\mathbf{r}) \, e^{i\mathbf{q}\mathbf{r}} d\mathbf{r}$ is the form factor of the atomic electron density oscillation harmonics, which can be expressed through the n-th order atom's polarizability. Equation (8.3) describes emission of n-photons by the polarization channel and m-photons by the static one, including all cross terms. However, due to a different character of the dependence on the momentum transmitted (in this case, for BR on a neutral atom), the static and polarization amplitudes weakly interfere, so after integrating over the IP scattering angles, the cross section of the process is split into the sum of two terms: one of them is static, another is "purely" polarizational. The static term is given by Eqn. (8.1) and the polarization one is determined by the expression [92]

$$\sigma_{\rm PB}^{(s)} = \frac{4\,\pi\,P_s^2\,\sin^2\gamma}{p^2}\,\ln\left(\frac{p_a}{|p - p'|}\right)\,, \quad \mathbf{P}_s = \hat{\chi}^{(s)}\,\mathbf{E}_0^s\,. \tag{8.4}$$

Here the non-linear susceptibility tensor $\hat{\chi}^{(s)}$ is introduced; γ is the angle between the initial momentum of the electron and the field vector.

Equations (8.3) and (8.4) were used in [92] to calculate the spectral cross section of two-photon bremsstrahlung effect for an electron with energy of 100 eV on the xenon atom near the resonance transition $5p^6\,{}^1S_0 - 6p$ (the resonance wavelength $\lambda_0 = 126$ nm). In this case only one photon is emitted by the polarization channel, because the even-order susceptibility for a free atom is zero. Comparison of the derived spectral dependence with the cross section accounting for only the static channel suggests that polarization effects, primarily contributing near the resonance frequency, are quite significant at large off-resonance (or order of 10%) frequencies, too.

Equation (8.3) correctly takes into account polarization effects in multiphoton BR if interaction of the external electromagnetic field with the atom – target is weak. Then the description of the process by perturbation theory can be applied, leading to the possibility of the cross section of the process to be expressed through the non-linear susceptibility of atom. For strong interaction of atom with radiation — for example, for near-resonance BR — the usage of Eqn. (8.3) could be insufficient.

Multiphoton bremsstrahlung effect in a strong near-resonance laser field with account of polarization effects of the target's core outside the applicability frames of Eqn. (8.3) was studied in [93] using the method of specified quantum/classical current [94]. A universal description for multiphoton induced bremsstrahlung was obtained including both static and polarization channels. An essential limitation of treatment [93] is the dipole approximation for interaction between the IP and the target's electronic core. Notice that the accuracy of this approximation increases with the target's ionization degree.

In order to describe polarization induced BR, in paper [93] the methods of equivalent Fermi photons and Kramers electrodynamics [46, 47] were generalized on multiphoton processes.

The probability of the s-photon process including both channels at a given electron scattering has the form [93]

$$W_{\sum}(s) = J_s^2 \{\rho \, |1 - \delta|\}, \qquad (8.5)$$

where, as before, $\rho = \mathbf{q}\,\mathbf{E}_0 / \omega^2$ and the function

$$\delta = \frac{\omega^2 \alpha(\omega, \mathbf{E}_0)}{Z_{\mathrm{i}}} \qquad (8.6)$$

is the ratio of the polarization and static channel amplitudes. In the near-resonance case we have

$$\delta = \left(\frac{\omega^2}{Z_{\mathrm{i}}}\right) \frac{d^2 \, \mathrm{sgn}(\omega - \omega_0)}{\Omega_{\mathrm{R}}}. \qquad (8.7)$$

Here $\Omega_{\mathrm{R}} = \sqrt{(\mathbf{d}\mathbf{E}_0)^2 + (\omega - \omega_0)^2}$ is the generalized Rabi frequency, ω_0, $\mathbf{d}$ the proper frequency and the dipole moment of the near-resonance transition, respectively, and Z_{i} is the ion – target charge.

Figure 22 shows calculations [93] for differential cross sections of bremsstrahlung absorption as a function of the strength of a linearly polarized laser field in the electron scattering on ion N^{4+} for the single-photon (curve 1) and two-photon (curve 2) processes. In the same Figure, the curves indicating contributions due to only static channel for single-photon (curve 3) and two-photon (curve 4) processes are also presented.

A generalization of calculations of the multi-photon BR with account of the polarization channel on the case of relativistic laser field intensities is given in paper [95].

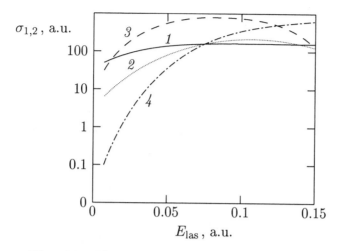

Fig. 22. The cross sections of the single-photon and two-photon bremsstrahlung absorption in electron scattering on the N^{4+} ion as a function of the laser field strength calculated in [93] with account for the polarization channel (curves *1, 2*) and in the static approximation (curves *3, 4*).

8.2.   Polarization – interference effects in collisions of an electron with atoms and ions in a near-resonance laser field

Consider polarization effects that appear in the inelastic electron scattering on an atom or ion with the core in a near-resonance electromagnetic field. The detuning of the external field frequency ω off the proper frequency of the target's electronic core ω_0 ($\Delta = \omega - \omega_0$) is assumed, on the one hand, to be sufficiently small so that the contribution due to other transitions to the process can be disregarded. On the other hand, the modulus of the detuning must be sufficiently large ($\Delta| \gg \gamma$, where γ is the near-resonance transition line width) in order that the actual excitation of the upper electronic state and the related radiative decay can be neglected. Indeed, the probability of the process with a real excitation of a discrete level is proportional to the imaginary part of target's polarizability. Thus the second term of Eqn. (6.3) implies that in the case of the near-resonance external field frequency the cross section with real excitation reads $\sigma^{(\text{Im})} \propto \gamma^3/(\Delta^2 + \gamma^2)^2$, while the cross section of interest is $\sigma^{(\text{Re})} \propto \gamma \Delta^2/(\Delta^2 + \gamma^2)^2$. As a result, the relative contribution of the cross section with the real level excitation is $(\gamma/\Delta)^2$, i.e. is negligible for large detunings. Note that this conclusion agrees with an analysis of atom excitation in the line wing by light pulses of short duration [96]. As was shown in paper [96], the "incoherent" part of the excited state population that remains after the external field turning off, is inversely proportional to the fourth power of the frequency detuning, while the "coherent" component that traces the instantaneous value of the radiative intensity, is inversely proportional to the second power of Δ.

As the contributions due to static and polarization channel are summed up in the BR amplitude, a cross-channel interference should be manifest in the total BR. However, in the cross section of PBR of fast electrons on neutral atoms, integrated over the incident particle scattering angle, the role of this interference is small. Indeed, in that case a spatial separation of the formation regions of static and polarization channels takes place. The static channel forms at small distances between the incident particle and the target's nucleus $r < R_0$, where the atomic electric field is strong and acceleration of an electron is large. The polarization channel, in contrast, is essential at large distances $r > R_0$, since at small r the polarization potential is low.

The cross-channel interference was ignored in considering laser radiation absorption in the optical gas break-down with account for the polarization mechanism and multiphoton effects [97, 98]. Here using the density matrix formalism, a quantum kinetic equation was derived for the electron energy gain in its scattering on atoms in resonance with the external field. Different regimes of the laser field inclusion were analyzed. The ratio of the polarization and static channels in terms of the inelastic electron scattering cross section on an atom (σ_{in}) and transport cross section (σ_{tr}) was found to be:

$$\xi = \frac{1}{8} \frac{\sigma_{\text{in}}}{\sigma_{\text{tr}}} \left(\frac{\omega}{\Delta}\right)^2 \frac{\omega}{\langle \varepsilon \rangle} f_{12}, \qquad (8.8)$$

Here f_{12} is the oscillator strength for the resonant transition; $\langle \varepsilon \rangle$ is the mean electron energy; $\Delta = \omega - \omega_0$ is the frequency detuning. For parameters used in [98], ratio (8.8) is $\xi \approx 10^3 - 10^7$. The estimation obtained in [98] for the threshold break-down intensity of cesium vapors for a relatively large detuning ($\Delta = 0.328$ eV), corresponding to the ruby laser energy and the minimal proper frequency of the transition in the cesium atom, gives $I_{br}^{theor} = 0.94 \cdot 10^9$ W/cm^2, which is in good agreement with experimental data $I_{br}^{exp} = 10^9$ W/cm^2 [99]. Note that here the probability ratio of the polarization and static channels is equal to $\xi \approx 85$, so that the main contribution to the process is exactly due to the polarization mechanism for the optical break-down.

The interference between the polarization and static mechanisms for the bremsstrahlung absorption disappears when averaging over the electron scattering angles [97, 98]. Nevertheless, even in the case of PBR of electrons on neutral targets, cross-channel interference effects can be manifest in the differential on the IP scattering angle cross section of the process.

The cross-channel interference in PBR was first calculated in paper [100] for the inverse bremsstrahlung effect as a function of the electron scattering angle on the hydrogen atom within the frames of the first Born approximation. The electromagnetic field frequency was $\omega = 0.3$ a.u., the initial electron momentum was p_i a.u.; in addition, the initial velocity vector was assumed to be perpendicular to the electric field strength vector. The process cross section was found to vanish for an electron scattering angle of 0.33 rad. This effect is due to a destructive interference between the static and polarization channels, whose amplitudes at given frequency and electron scattering angle have the opposite signs and close values. By increasing the electron scattering angle the static channel becomes dominant, while at small angles the polarization channel prevails. With the external field frequency approaching one of the target's eigenfrequencies, the scattering angle at which the process cross section vanishes increases because the polarization term amplitude increases.

In paper [101], two-photon free-free transitions in a laser field during the electron scattering on the hydrogen atom were considered with an account for the target's polarization within the frames of the first Born approximation. Spectral-angular relationships of the process were obtained for different external field vector polarizations with respect to the initial IP momentum and various collision energies. The calculation revealed an interference dip in the dependence of the cross section on the IP energy, due to destructive interference between the static and polarization channels. Similar interference dips were discovered in angular dependences for the two-photon absorption at a fixed IP energy. In [10] was also noted that inside the interference dips the cross section vanishes for low energies of the external field photons ($\omega < 6.8$ eV) when the channel amplitudes are real values. If the photon frequency is such that an imaginary part in the target's polarizability appears, the zero dip transforms into a shallow minimum.

The role of the cross-channel interference effects in the integral over the IP scattering angles cross section for the bremsstrahlung emission/absorption must increase with the degree of inelasticity of the process, i.e. when the energy of the emitted/absorbed photon becomes of the order of (or exceeds, in the case of the bremsstrahlung absorption) the initial IP energy. In this case, as follows from energy – momentum conservation, the spatial formation regions of the static and polarization channels start stronger overlapping. However, for interference effects to show up the channel amplitudes must be comparable, too. This occurs, for example, for BR of slow electrons on ions with a core under the conditions for the Kramers electrodynamics to be applicable [47], when the radiative process is spatially localized. If the effective radiation radius here exceeds the size of the target's electronic core, when the dipole approximation for the IP interaction with the electronic core is valid, the cross-channel interference can be also manifest in the integral over the electron scattering angle cross section for the bremsstrahlung emission/absorption.

Interference signatures in the bremsstrahlung spectral cross section are most prominent in the near-resonance region, which is evident from Eqn. (8.5) for the considered process probability at given electron scattering. For example, if the absolute values of the channel amplitudes match in the low frequency wing ($\delta = 1$), when the target's polarizability is positive, the probability of the total bremsstrahlung emission/absorption vanishes by the destructive cross-channel interference. Remarkably, as the external field crosses the resonance, the cross-channel interference changes the character, which is accounted for by the factor sgn $(\omega - \omega_0)$ in Eqn. (8.7). For $\omega > \omega_0$ the cross-channel interference becomes constructive and the total probability of the process exceeds its static value. Moreover, interference effects in the near-resonance case significantly depend on the electric field strength, as indicated by Eqn. (8.7) and the generalized Rabi frequency definition. By increasing the laser field strength E_0 the polarization amplitude decreases and correspondingly the interference dip frequency in the spectral cross section of the process approaches the transition eigenfrequency. This phenomenon is similar to the saturation effect in radiation absorption by a two-level system.

It is essential that in the dipole approximation on the IP interaction with target's electronic core we used in deriving Eqn. (8.5), the corresponding probability is independent of the electron scattering angle. So the interference feature described above remains in the integral over the IP scattering angle cross section of the process, unlike the case of fast particles. In fact, Eqn. (8.7) does not take into account contributions due to small distances to the bremsstrahlung emission/absorption probability, when IP penetrates into the electronic core of the target and its interaction with target's electrons becomes significantly non-dipole.

The role of penetration effects in the near-resonance BR was analyzed in paper [102] for electron scattering on multicharged ions with a core. Specific calculations were carried out for resonant transitions in lithium-like ions without ($\Delta n = 0$) and with ($\Delta n \neq 0$) the principle quantum number change. The IP penetration into the electronic core of the target was found to differently affect different Cartesian components (in the local frame) of the radiating dipole moment vector induced in the ionic core. For example, in the case $\Delta n = 0$, this influence proves strong for the x-component of the dipole moment and is much weaker for the y-component. This leads to a significant dependence of polarization – interference features of the bremsstrahlung effect on the angle α made by the initial IP velocity $\mathbf{v}_0$ with the electron field strength vector in the electromagnetic radiation. This dependence is determined by the difference of the D_x and D_y projections of the emitting dipole moment on the axes of the focal frame. In the Kramers high frequency limit ($\omega \gg v^3/Z_i$), the main contribution to the BR cross section is due to trajectories with high orbital eccentricities ($\varepsilon \leq 1$) [9], then $D_x \approx \cos^2 \alpha$ and $D_y \approx \sin^2 \alpha$. Thus for the parallel orientation of the field vector relative to the initial velocity vector ($\alpha = 0$), the main contribution to the cross section is due to the x-component of the dipole moment, and due to the y-component in the case of the perpendicular orientation ($\alpha = \pi/2$). But since the x-component of the dipole moment induced in the core strongly alters by the IP penetration in the core, and the y-component (at least for transitions with $\Delta n = 0$) changes much smaller, then for $\alpha = 0$, the IP penetration strongly suppresses the cross-channel interference in the integral cross section; if $\alpha = \pi/2$, this suppression is not that strong. This is illustrated by Fig. 23, which shows the ratio of the total cross section of the near-resonance BR in the electron scattering on the N^{4+} ion to the static channel cross section for two values of the angle α as a function of frequency detuning off the resonance. It is evident that for the parallel orientation, the interference dip in the spectral cross section disappears, while for the perpendicular orientation a relatively broad minimum emerges in the low frequency line wing due to destructive interference between the static and polarization channels of the process. In the high frequency wing, the relative difference of the cross sections for two values of the angle α is small. In this case the cross-channel interference has constructive character, so for the perpendicular orientation the cross section somewhat exceeds the value for the parallel orientation.

Polarization – interference effects in PBR are much more prominent in the differential over the IP scattering angle cross section of the process. Spectral-angular dependences for the near-resonance bremsstrahlung effect in a strongly inelastic electron scattering on ions with a core were analyzed in [103] for quantum IP motion. These dependences, averaged over the spin state, indicate that for the parallel orientation of the field (with respect to the initial IP velocity), the interference dip in the spectral cross section shifts from the low frequency wing to the high frequency one with an increase in the electron scattering angle. In the case of the perpendicular orientation of the field vector, the interference dip frequency always locates

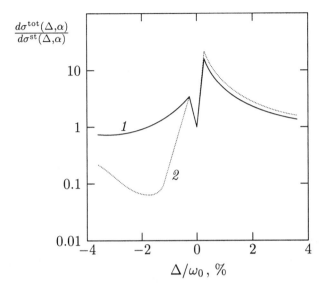

Fig. 23. The spectral dependences of the induced bremsstrahlung emission/absorption (normalized to the static value) from quasi-classic electron scattering on the N^{4+} ion for two angles α ($\alpha = 0$, curve *1*, $\alpha = \pi/2$, curve *2*) between the incident particle initial velocity ($v_i = 0.6$ a.u.) and the external radiation electric field vector ($E_0 = 10^{-3}$ a.u.), for the external field frequency at the proper frequency of the near-resonant transition in the ionic core without the principal quantum number change ($2s - 2p$) [102].

in the low frequency wing, by shifting toward the line center with an increase in the IP scattering angle.

Within the frames of quantum treatment, the basic formula for the cross section of inelastic electron scattering (from the state with the initial momentum $\mathbf{p}_i$ to the state with the final momentum $\mathbf{p}_f$) inside the solid angle $d\Omega_f$ on an ion, assisted by an external field of amplitude $\mathbf{E}_0$ and frequency ω, can be presented in the form (using throughout atomic units $\hbar = m = e = 1$):

$$d\sigma(\mathbf{p}_f) = \frac{1}{16\pi^2} |(M_{fi})|^2 \frac{p_f}{p_i} d\Omega_f, \qquad (8.9)$$

where

$$M_{\mathrm{fi}}(\omega) = \langle \mathbf{p}_{\mathrm{f}} | \, \frac{\mathbf{r}}{r} \, (V_{\mathrm{st}}(r) + V_{\mathrm{pol}}(r, \omega, E_0)) \, | \mathbf{p}_{\mathrm{i}} \rangle \, \frac{E_0}{\omega^2} \qquad (8.10)$$

is the matrix element of the perturbation operator of the IP motion in the static field of the target and in the field of the target's induced dipole moment. Here $|\mathbf{p}_{\mathrm{i,f}}\rangle$ are the IP wave functions in the central field of the ion normalized to the unitary flux at infinity, $p_{\mathrm{f}} = \sqrt{p_{\mathrm{i}}^2 \pm 2\omega}$ is the modulus of the IP final momentum, the positive sign relating to the absorption and the negative sign to the emission of the photon during scattering.

Expression (8.10) is obtained by neglecting exchange effects, whose contribution to the cross section will be discussed below. The function $V_{\mathrm{st}}(r)$ is the modulus of the ion–target's static field strength at the IP location site and $V_{\mathrm{pol}}(r, \omega, E_0)$ is linked to the dynamic polarizability of ion's electronic shell at the external field frequency ω. For the near-resonance case considered here ($|\omega - \omega_0| \ll \omega_0$) has the form

$$V_{\mathrm{pol}}(r, \omega, E_0) = \mathrm{sgn}(\Delta) \, \frac{d_0}{3} \, \frac{\omega^2}{\sqrt{\Delta^2 + (d_0 E_0)^2/3}} \, V_{ns,n'p}(r) \qquad (8.11)$$

where $d_0 = \langle ns || d || n'p \rangle$ is the reduced matrix element of the dipole moment of the transition, $V_{ns,n'p}(r)$ is the reduced matrix element of the IP interaction potential with the near-resonance transition $ns \to n'p$ in the ion's core which is

$$V_{ns,n'p}(r) = \left\langle ns \left\| \theta(r - r_b) r_b \middle/ r^2 + \theta(r_b - r) r \middle/ r_b^2 \right\| n'p \right\rangle. \qquad (8.12)$$

Here $\mathbf{r}_0$ is the radius–vector of ion's bound electron and $\theta(x)$ is the Heaviside function. Everywhere we consider the ground s-state of the ionic core. In calculating functions $V_{\mathrm{st}}(r)$ and $V_{ns,n'p}(r)$ for the outer electron of ion's core, the wave function of the model potential was used. For the internal $2s$ electrons Slater's wave function were employed.

In this treatment (unlike previous quantum calculations [93, 97]), the interference of contributions due to static and polarization channels in non-dipole IP interaction with the radiative transition inside the ion's core has been consistently taken into account.

The non-dipole character is important if the contribution from small distances (of the order of ion size) exceeds or is comparable with that from large distances, occurring in the case of a sufficiently strong scattering inelasticity as considered below.

The IP wave functions can be computed using the approximation of a given quantum current of IP by expanding in spherical harmonics corresponding to particular values of the orbital momentum l. The external electromagnetic field is assumed to be sufficiently weak so that the wave functions of the scattered electron continuum can be found by solving the corresponding Schrödinger equation in the central field of the ion-target. The calculation was carried out for lithium-like ions in the ground state and for the external radiation frequencies near the frequency of the transition without changing the principle quantum number.

The radial wave functions of the IP continuum were normalized in agreement with the asymptotic

$$u(r \to \infty, l, p) \to \frac{2}{r} \sin\left(pr + \frac{Z_i}{p} \ln(2pr) - \frac{\pi}{2} l + \delta(l, p)\right) . \quad (8.13)$$

Here $\delta(l,p) = \delta^{\mathrm{C}}(l,p) + \Delta\delta(l,p)$ is the total phase shift equal to the sum of the Coulomb $\delta^{\mathrm{C}}(l,p)$ and non-Coulomb $\Delta\delta(l,p)$ phase shifts, with the latter being calculated using the formula [74]:

$$\sin(\Delta\delta(l,p)) = \frac{1}{2p} \int_0^\infty \left(\frac{Z_i}{r} - |V_{\mathrm{st}}(r)|\, r\right) u(r, l, p)\, u^{\mathrm{Coul}}(r, l, p)\, r^2\, dr ,$$

$$(8.14)$$

where $u^{\mathrm{Coul}}(r, l, p)$ is the solution of the radial Schrödinger equation with Coulomb potential.

As a result, the integral and differential cross sections of the inelastic scattering are represented as the sum (over the orbital quantum number) of terms containing radial matrix elements of the total perturbation potential of the IP motion ($V_{\mathrm{st}}(r) + V_{\mathrm{pol}}(r, \omega, E_0)$):

$$R_{l,l\pm1} = \langle u(r, l, p_{\mathrm{i}})|V_{\mathrm{st}}(r) + V_{\mathrm{pol}}(r, \omega, E_0)|u(r, l \pm 1, p_{\mathrm{f}})\rangle \quad (8.15)$$

and the corresponding phase shifts of the IP wave functions.

After integrating over the IP radius-vector directions, Eqns (8.9) and (8.10) yield

$$d\sigma(\Omega_f) = \frac{1}{16 p_f p_i^3} \left(\frac{E_0}{\omega^2}\right)^2 \left|\sum_{l=0}^{\infty} S_l^{fi}(\Omega_f)\right|^2 d\Omega_f . \qquad (8.16)$$

For the parallel polarization ($\mathbf{p}_i \parallel \mathbf{E}_0$) terms S_l read

$$S_l^{fi,par}(\theta) = a_l \left[\begin{array}{c} P_{l+1}(\cos\theta)R_{l,l+1}e^{i(\delta(l,p_i)+\delta(l+1,p_f))} - \\ P_l(\cos\theta)R_{l+1,l}e^{i(\delta(l+1,p_i)+\delta(l,p_f))} \end{array}\right] \qquad (8.17)$$

where $P_l(\cos\theta)$ are the Legendre polynomials, θ is the IP scattering angle, $a_l = l + 1$.

In a similar way, for the perpendicular polarization of the external field ($\mathbf{p}_i \perp \mathbf{E}_0$)

$$\begin{aligned} S_l^{fi,per} = e^{i\delta(l+1,p_f)} b_l [Y_{l+1,1} + Y_{l+1,-1}] \\ \times \left\{e^{i\delta(l,p_i)}R_{l,l+1} + e^{i\delta(l+2,p_i)}R_{l+1,l}\right\}, \end{aligned} \qquad (8.18)$$

where $b_l = \sqrt{\frac{\pi(l+1)(l+2)}{2l+3}}$, $Y_{nm}(\Omega_f)$ are spherical functions.

For the strongly inelastic scattering considered here the most contribution to the cross section is due to small distances to ion's core. In calculating the radial matrix elements, the exchange processes in the polarization channel were included for two possible total spins of the "IP+ion's core" system.

The results of the spectral cross section calculations normalized to the static cross section and average over possible values of the total spin of the colliding particles are shown in Fig. 24(a,b) for different IP inelastic scattering angles and two external field polarizations. The incident electron energy is 1 Ry and the external field strength is $E_0 = 10^{-3}$ a.u. with the frequency near the resonance transition $2s - 2p$ ($\hbar\omega = 10$ eV) in the N^{4+} ion core.

Figure 24(a) indicates that taking into account the non-dipole character of the IP interaction with a radiative transition inside an ion's core in the case of the external field parallel polarization leads to a significant dependence of the spectral cross section minimum, caused by destructive interference between the polarization and the

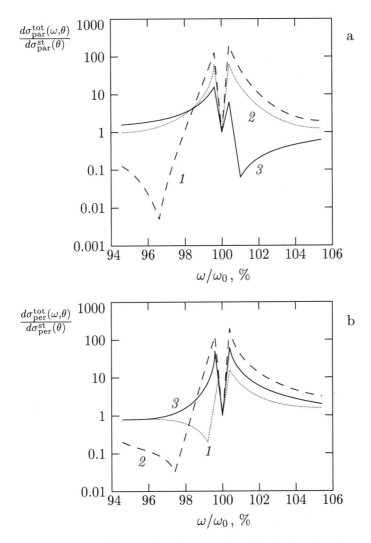

Fig. 24. Spectral cross sections for inelastic electron scattering (absorption) on the N^{4+} ion, averaged over the spin state of the system of colliding particles, for different IP scattering angles: $\theta = 57°$ (curve *1*), 120° (curve *2*), 140° (curve *3*), and normalized to the corresponding static cross sections. Quantum calculation: a — the parallel polarization of the external field $\mathbf{p}_i \parallel \mathbf{E}_0$, b — the perpendicular polarization of the external field $\mathbf{p}_i \perp \mathbf{E}_0$.

static channel, on the electron scattering angle. For small IP scattering angles ($\theta < 90°$), the minimum corresponds to negative detunings ($\omega < \omega_0$), while for large angles ($\theta > 140°$) — to positive ($\omega > \omega_0$). There is a small range of angles ($\theta \approx 120°$) inside which the minimum is absent. Correspondingly, an "inversion" of the spectral cross section line form asymmetry occurs as the IP scattering angle increases.

In the case of perpendicular polarization ($\mathbf{p}_i \perp \mathbf{E}_0$), the interference minimum for all IP scattering angles lies within the range of negative detunings of the external field frequency off the resonance inside the ion core (Fig. 24(b)), by shifting toward the line center with an increase of the scattering angle and vanishing at $\theta \approx 180°$.

The above features of the differential over the IP scattering angle cross section are due to the dependence of the radial matrix element of the non-dipole polarization interaction on the quantum number of the IP orbital momentum. This matrix element changes the sign in passing from large to small momentum values. The role of these small momenta is especially large for the parallel polarization of the external field, since in this case the contribution due to the IP orbit part near the classical turn-off point increases. In contrast, for the perpendicular polarization, there is an increase in the role of large IP-ion distances (and, respectively, large orbital momenta) where the IP acceleration proves parallel to the external field strength vector.

The integral over IP scattering angle spectral cross section for induced inelastic scattering (for the parallel orientation of the external field strength vector) is obtained in the form

$$\int d\sigma_{\mathrm{P_f}}^{\mathrm{par}} = \frac{\pi}{4} \left(\frac{E_0}{\omega^2}\right)^2 \frac{1}{p_f p_i^2} \sum_{l=0}^{\infty} \frac{(l+1)^2}{2l+3} T_l^{\mathrm{par}}(p_i, p_f), \qquad (8.19)$$

in which the quantity T_l^{par} is defined by the combination of radial matrix elements:

$$T_l^{\mathrm{par}}(p_i, p_f) = R_{l,l+1}^2 + \frac{2l+3}{2l+1} R_{l+1,l}^2$$
$$- 2\frac{l+2}{l+1} R_{l,l+1}^2 R_{l+2,l+1}^2 \cos\left(\delta(l,p_i) - \delta(l+2,p_i)\right).$$

Averaging Eqn. (8.19) over the external field polarization (or, which is equivalent, over the initial IP momentum), we get

$$\int d\sigma_{\mathbf{P}_{\mathrm{f}}}^{\mathrm{aver}} = \frac{\pi}{4} \left(\frac{E_0}{\omega^2}\right)^2 \frac{1}{p_{\mathrm{f}} p_{\mathrm{i}}^2} \sum_{l=0}^{\infty} \frac{(l+1)}{3} (R_{l,l+1}^2 + R_{l+1,l}^2). \quad (8.20)$$

For an arbitrary orientation of vector $\mathbf{E}_0$ relative to vector $\mathbf{p}_i$, we find

$$\int d\sigma_{\mathbf{P}_{\mathrm{f}}}(\alpha) = \frac{3}{2} \sin^2 \alpha \left\{ \int d\sigma_{\mathbf{P}_{\mathrm{f}}}^{\mathrm{aver}} \right\} + \frac{3 \cos^2 \alpha - 1}{2} \left\{ \int d\sigma_{\mathbf{P}_{\mathrm{f}}}^{\mathrm{par}} \right\},$$
$$(8.21)$$

where α is the angle between vectors $\mathbf{p}_i$ and $\mathbf{E}_0$. It is important that the quantum calculation allows to establish the difference between the photon absorption and emission cross section in the IP scattering in the external field, essential for estimating the energy exchange between plasma and radiation. However, as the above calculations imply, the relative value of the corresponding difference cross section (absorption minus emission of a photon) is maximal for a directional IP motion and strongly diminishes when averaging over the angle α between the field vector and the IP initial velocity vector.

The integrals over the IP scattering angle spectral cross section for emission and absorption of photons are shown in Fig. 25(a,b) for various external field polarizations. The electron scattering with an energy of 11 eV on the N^{4+} ion is considered for a field in resonance with the $2s - 2p$ transition in the ion core, with $E_0 = 10^{-3}$ a.u. The results exhibit qualitative agreement with the previous quasi-classical treatment, according to which interference effects in the integral cross section for inelastic scattering are pronounced mostly for the perpendicular polarization of the external field. At the same time, the spectral dip in the quantum theory proves more smoothed and less deep compared to classical calculations.

For large detunings off the resonance where static channel prevails, the sign of the difference cross section is consistent with calculations made within the first Born approximation [104]: absorption exceeds emission for the perpendicular polarization, and *vice versa*. With decreasing the frequency detuning, the difference cross section starts being mostly determined by the cross-channel interference. As

V. A. Astapenko et al

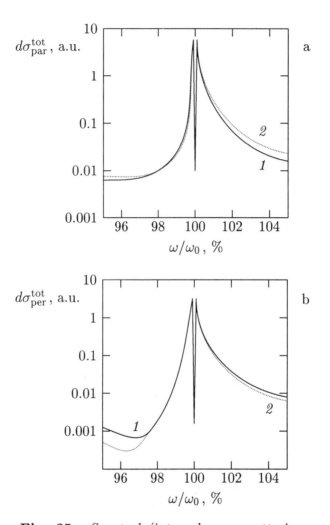

Fig. 25. Spectral (integral over scattering angles) absorption (curve *1*) and emission (curve *2*)
cross sections in electron scattering ($p_i = 0.9$ a.u.)
on the lithium-like nitrogen in the external field
($E_0 = 10^{-3}$ a.u.), which is near-resonant to the
$2s - 2p$ transition in the core. Quantum calculation: a — the parallel polarization of the external
field $p_i \parallel E_0$, b — the perpendicular polarization
of the external field $p_i \perp E_0$.

was already mentioned, this is mostly pronounced in the integral scattering cross section for the perpendicular polarization of the external field. For example, in the frequency range where destructive interference dominates, absorption notably exceeds emission for the perpendicular polarization, as is evidenced from Fig. 25(b), since the spectral minimum of scattering with photon emission is shifted towards large detunings compared to the case with absorption. However, the same shift causes the emission cross section to start exceeding the absorption one with a decrease in the frequency detuning value at $\omega < \omega_0$. For $\omega > \omega_0$ ($\mathbf{p}_i \perp \mathbf{E}_0$) the situation is opposite.

For the parallel polarization ($\mathbf{p}_i \parallel \mathbf{E}_0$), the difference cross section value inside the spectral interval where interference effects are significant is much less (Fig. 25(a)) and has the opposite sign: absorption exceed emission for sufficiently small absolute values of the negative detunings ($\omega < \omega_0$), and *vice versa* for $\omega > \omega_0$.

The calculation carried out within the frames of the model considered here indicates that the averaged over the angle α and the total spin difference cross section for inelastic scattering amounts to an appreciable value near the resonance, where the process mainly occurs via polarization channel. The sign of the difference cross section strongly depends on the IP energy: immediately at the threshold (for the IP scattering with emission of a photon), emission exceeds absorption, but already for small IP energy excesses over the threshold, the situation becomes reversed. Thus the quantum treatment considered above confirms conclusions of quasi-classical calculations about the important role polarization effects play in the near-resonance bremsstrahlung in a strongly inelastic electron scattering. Moreover, the quantum approach allows the differential over the IP scattering BR cross section to be correctly calculated, the processes with photon emission and absorption to be distinguished, and quasi-classical results to be completed with an account for spin effects.

9. Polarization radiation, Compton scattering and collisional ionization. Cross section relationships, similarity laws, new ionization cross section data

Now we consider how the atomic plasma model can be applied to describe collisional-radiative processes accompanied by target's ionization. These include collisional ionization of atoms (CIA) and PBR with target's ionization or incoherent PBR. In these processes, the IP momentum – energy excess is transmitted not to the target as a whole, but to one electron being ionized. In classical paper [22], in describing atom excitation and ionization by charged particles E. Fermi used the analogy between the charged particle's field and radiation equivalent to it (the equivalent photon method). Here in order to calculate the incoherent PBR and CIA cross sections, we shall use an analogy with X-ray scattering. Moreover, the statistical model of atom is applied to establish an approximate scaling for the atomic Compton profile, allowing for a universal description (for all nuclear charges) of the X-ray scattering cross section by atoms and related processes of incoherent PBR and CIA.

9.1. Approximate scaling of the atomic Compton profile

In the case of X-ray scattering by atomic electrons, their coupling with the nucleus is known to significantly modify (see, for example, [23]) the process cross section. As a result, the unique correspondence between the scattering angle and the frequency shift of scattered radiation smears out. Thus, each scattering angle corresponds to some frequency distribution of scattered X-rays centered on the frequency determined by the well-known Compton formula for free electron. The width of this distribution measures the coupling of the electron with the nucleus. The corresponding frequency-angular dependence of the cross section can be most simply described within the frames of the so-called momentum approximation when the interaction time of radiation with matter is small and the scattering process occurs both on free and bound electron in a similar way. In this case, the scattering (on a bound electron) is treated like on a free electron but with account for the initial distribution over momentum. Then the effect of the electron – nucleus coupling on scattering can be described using the Compton scattering profile (CP), $J(Q)$, containing information on the momentum distribution of atomic electrons. This quantity for the nl-th atomic subshell is given by the formula [105]

$$J_{nl}(Q) = \frac{1}{2} \int\limits_{Q}^{\infty} |R_{nl}(p)|^2 p \, dp \,. \tag{9.1}$$

Here $R_{nl}(p)$ is the spatial Fourier-image of the normalized radial wave function of the nl-state. The argument of the function $J(Q)$ represents the projection of the initial atomic electron momentum in the direction determined by the X-ray wave vector change. As the Compton scattering is an incoherent process, the atomic CP is equal to the sum of CP of all electronic subshells.

The atomic CP in the non-relativistic momentum approximation considered here enters as a multiplier into the expression for the Compton cross section and hence determines its frequency-angular dependence. There are extensive numerical tables for CP of all shells of all elements [105]. An analysis of these data indicates that a similarity law can be established for the atomic CP using the dimensionless transmitted momentum $\tilde{Q} = Q \, r_{\mathrm{TF}}$ and the reduced Compton profile

$$\tilde{J}(\tilde{Q}) = \frac{1}{Z \, r_{\mathrm{TF}}} \, J\left(\frac{\tilde{Q}}{r_{\mathrm{TF}}}\right) \,. \tag{9.2}$$

By constructing the reduced CP $\tilde{J}$ for different multielectron atoms as a function of the dimensionless momentum using data [105], for not very large arguments (inside the range $\tilde{Q} \leq 10)$), we discover that the corresponding curves resemble each other closely and can be uniquely approximated by the relationship

$$\tilde{J}_{\mathrm{scal}}(\tilde{Q}) = \frac{0.8}{1 + \left(\tilde{Q}\big/0.4\right)^2} \,, \tag{9.3}$$

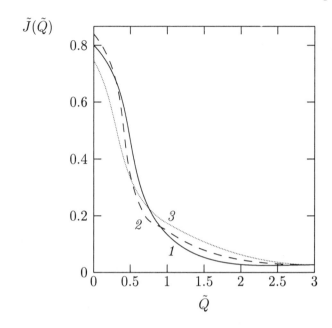

Fig. 26. The dependence of the reduced Comp-
ton profile $\tilde{J}(\tilde{Q})$ on the reduced momentum $\tilde{Q}$:
1 — universal scaling (9.3), *2* — calculations for
the argon atom, *3* — calculations for the krypton
atom [105].

This dependence is illustrated in Fig. 26 which displays the reduced CP for argon and krypton, calculated using Eqn. (9.2) and tables [105] together with similarity law (9.3). Clearly, scaling (9.3) rea-
sonably well describes the Hartree – Fock CP values [105] within the
given range of the transmitted momenta. Making use of Eqns (9.2)
and (9.3), a universal expression for the Compton scattering cross
section on atom can be derived within the frames of the so-called
momentum approximation [23] that describes the process for all nuclear charges in a universal way. In what follow we shall use approximate scaling (9.3) to derive the similarity law for the cross section
of collisional ionization of atoms and the universal formula for the
incoherent PBR cross section.

9.2. Collisional ionization of atoms. The cross section calculation in the Born – Compton approximation

Ionization of atoms resulted from collisions with charged particles is the fundamental process that plays an important role in different fields of physics and many technical applications. There are many ways to calculate this process cross section based on various models, approximations and semi-empirical formulas (see for example [106 – 109]). The results of such approximate calculations (using the minimum number of parameters) in some cases demonstrate good agreement with experiments [110], in other cases significant discrepancy is evident. Here we propose a simple and novel method to calculate the collisional ionization cross section that generalizes known methods on the situations with the largest disagreement with experimental measurements. The method proposed is named the Born – Compton (BC) approach as it is based, in addition to using the Born approximation, on describing the dynamics of target's electrons using the X-ray scattering Compton profile (9.1) and its approximate scaling (9.3).

Using a standard procedure within the frames of the first Born approximation, it is not difficult to obtain the following expressions for the differential cross section of ionization from the nl-th atomic subshell (using we use atomic units throughout):

$$d\sigma_{nl} = \frac{4\,Z_{\mathrm{pr}}^2}{v}\, S_{nl}(q)\, \frac{d\mathbf{q}}{\mathbf{q}^4}\,. \tag{9.4}$$

Here Z_{pr} is the IP charge, v its velocity $q = (q^0 = E_{\mathrm{f}} - E_{\mathrm{i}},\ \mathbf{q} = \mathbf{p}_{\mathrm{f}} - \mathbf{p}_{\mathrm{i}})$ is the IP energy – momentum change, and S_{nl} is the dynamic form factor (DFF) of the subshell that is the spatial Fourier-image of the autocorrelation function of the density of atomic electrons.

Making use of the plane wave approximation for wave functions of the target's continuum spectrum, DFF can be casted in the form

$$S_{nl}(q) = \int \frac{d\mathbf{p}}{4\,\pi}\, \delta\left(q^0 + \frac{(\mathbf{p} - \mathbf{q})^2}{2} - \varepsilon_{nl}\right)\, |R_{nl}(p)|^2\,. \tag{9.5}$$

Here ε_{nl} is the binding energy of the subshell and $R_{nl}(p)$ is the radial wave function of the subshell in the momentum representation determined by the expression

$$R_{nl}(p) = \sqrt{\frac{2}{\pi}} \int\limits_0^\infty R_{nl}(r)\, j_l(p\,r)\, r^2 dr\,, \qquad (9.6)$$

where $j_l(p\,r)$ is the spherical Bessel function of the first kind and $R_{nl}(r)$ is the normalized wave function.

Here we employ the momentum approximation, within which bound electrons are considered as quasi-free with the corresponding momentum – energy connection determined by the equality

$$p^2 \big/ 2 = \varepsilon_{nl}\,. \qquad (9.7)$$

Then from general formula (9.5) the following representation of DFF through the electronic subshell CP $J_{nl}(Q)$ can be obtained:

$$S_{nl}^{(IA)}(q) = \frac{1}{|\mathbf{q}|}\, J_{nl}\left(Q = -\frac{q^0 + \mathbf{q}^2/2}{|\mathbf{q}|} \right) \qquad (9.8)$$

and then using Eqns (9.4) – (9.8) the differential over the displaced electron's energy W cross section of the collisional ionization can be found in the form

$$\frac{d\sigma_{nl}(w,\,x)}{dw} = \frac{\sqrt{2}\,\pi\,Z^2}{I_{nl}^{3/2}} \frac{1}{x} \int\limits_{\sqrt{x}-\sqrt{x-w-1}}^{\sqrt{x}+\sqrt{x-w-1}} J_{nl}\left(\sqrt{\frac{I_{nl}}{2}}\, \frac{w+1-t^2}{t} \right) \frac{dt}{t^4}\,. \qquad (9.9)$$

Here the dimensionless variables $x = E_i/I_{nl}$, $w = W/I_{nl}$ have been introduced, where I_{nl} is the ionization potential of the electronic subshell.

Equation (9.9) yields the total ionization cross section from atom's electronic subshell

$$\sigma_{nl}(x) = \frac{\sqrt{2}\,\pi\,Z^2}{I_{nl}^{3/2}} \frac{1}{x} \int\limits_1^{y_m} dy \int\limits_{\sqrt{x}-\sqrt{x-y}}^{\sqrt{x}+\sqrt{x-y}} J_{nl}\left(\sqrt{\frac{I_{nl}}{2}}\, \frac{y-t^2}{t} \right) \frac{dt}{t^4}\,. \qquad (9.10)$$

In equation (9.10) the maximum dimensionless transmitted energy y_m is introduced. Below we discuss specific forms of this energy.

Summing up Eqn. (9.10) over all electronic subshells, we ultimately arrive at

$$\sigma_i(E_i) = \sum_{nl} N_{nl}\,\sigma_{nl}\,(E_i/I_{nl})\,\theta(E_i - I_{nl})\,. \qquad (9.11)$$

Here N_{nl} is the number of the equivalent electrons in the nl-th subshell, $\theta(x)$ is the Heaviside step function.

The result (9.10) and (9.11) is appropriate to name the Born – Compton (BC) approximation as it express the collisional ionization cross section in terms of the Compton profile, i.e. in terms of the same target's characteristic that describes the coupling of atomic electrons with the nucleus in the case of X-ray Compton scattering.

It is interesting to note that Eqn. (9.10) in the limit of free electrons is transformed to the well-known Thompson formula for collisional ionization of target. Indeed, in this case CP can be written in the form

$$J(Q) = \delta(Q)\,. \qquad (9.12)$$

Here $\delta(x)$ is Dirac's delta-function. Substituting Eqn. (9.12) into (9.10) and making elementary integration, we arrive at the Thompson cross section:

$$\sigma^{T\,\mathrm{hom}}(x) = \frac{\pi}{I^2}\,\frac{x-1}{x^2}\,. \qquad (9.13)$$

Using Eqns (9.10) and (9.11) and data from Tables [105], we can calculate the collisional ionization cross section; however, in the subsequent calculations we shall use scaling (9.3). The expression for CP that follows from Eqn. (9.3) can be recast through the electronic subshell ionization potential in the form

$$J^{(\mathrm{Sc})}(Q) = \frac{2.5}{\pi\,\sqrt{2\,I}}\,\frac{1}{1 + (Q^2/0.8\,I)}\,. \qquad (9.14)$$

Substituting Eqn. (9.14) into (9.10), we obtain the standard collisional ionization cross section expressed via the similarity function in the form:

$$\sigma_{nl}\,(x = E/I_{nl}) = \frac{\pi\,Z_{\mathrm{pr}}^2}{I_{nl}^2}\,f(x) \qquad (9.15)$$

where

$$f(x) = f^{(\mathrm{BC})}(x) = \frac{2.5}{\pi} \frac{1}{x} \int\limits_{1}^{y_\mathrm{m}} dy \int\limits_{\sqrt{x}-\sqrt{x-y}}^{\sqrt{x}+\sqrt{x-y}} \frac{dt}{t^2 \left[t^2 + (y - t^2)^2/0.64 \right]}.$$

(9.16)

The maximum dimensionless transmitted energy can be understood in two ways: (a) $y_\mathrm{m} = x$; (b) $y_\mathrm{m} = (x+1)/2$. Case (a) relates to collision of identical particles; case (b), according to paper [111], corresponds to ignoring the exchange within the 1st Born approximation, when the total spin of the collision particle system is not specified. Thus, depending on the choice of the upper limit in the external integral of equality (9.16), there appear two modifications of the Born–Compton approximation (a) and (b). For comparison, we show here other forms of the similarity function, such as Gryzinski's formula [106]:

$$f^{(\mathrm{Gryz})}(x) = \frac{1}{x} \left(\frac{x-1}{x+1} \right)^{3/2} \left[1 + \frac{2}{3} \left(1 - \frac{1}{2\,x} \right) \ln \left(2.7 + \sqrt{x-1} \right) \right]$$

(9.17)

and Eletskij–Smirnov's formula [108]:

$$f^{(\mathrm{ES})}(x) = \frac{10\,(x-1)}{\pi\,x\,(x+8)}$$

(9.18)

and also the and also the biparametric function of the BEB (binary-encounter-Bethe) approximation [109]:

$$f^{(\mathrm{BEB})}(x, u) = \frac{1}{1+x+u} \left[\frac{\ln(x)}{2} \left(1 - \frac{1}{x^2} \right) + 1 - \frac{1}{x} - \frac{\ln(x)}{1+x} \right].$$

(9.19)

Here u is the dimensionless mean kinetic energy of electrons normalized on the ionization potential of the given subshell. Note that relationships (9.18) and (9.19) are virtually coincident for $u = 0.6$.

The double integral in the definition of Born–Compton similarity function (9.16) can be approximated to a good accuracy by the following analytical formulas:

$$f_{\text{fit}}^{(BC(a))}(x) = \frac{0.51\,(x-1)}{x^{3/2} - 1.234\,x + 1.273\,\sqrt{x}}, \qquad (9.20a)$$

$$f_{\text{fit}}^{(BC(b))}(x) = \frac{0.514\,(x-1)}{x^{3/2} - 1.224\,x + 1.663\,\sqrt{x}} \qquad (9.20b)$$

which are convenient to use to evaluate the collisional ionization cross section in the Born–Compton approximation. All similarity functions quoted above are shown in Fig. 27. It is evident that the maximum of the Born–Compton function is shifted toward low energies compared to relationships (9.17)–(9.19); the function value at the maximum is larger, which is especially prominent for modification (9.20b). Indeed, the value and location of maxima in the BC approximation are, respectively: $f_{\text{max}}^{(BC(a))} = 0.277$, $x_{\text{max}}^{(BC(a))} = 3.3$, $f_{\text{max}}^{(BC(b))} = 0.24$, $x_{\text{max}}^{(BC(b))} = 3.85$, while for the semi-empirical Eletskij–Smirnov's function: $f_{\text{max}}^{(ES)} = 0.2$, $x_{\text{max}}^{(ES)} = 4$.

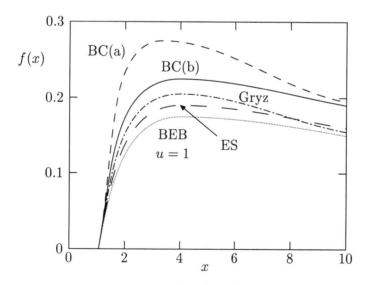

Fig. 27. The similarity functions $f(x)$ for the collisional ionization cross sections calculated for different models as a function of the incident particle reduced energy x (see the text).

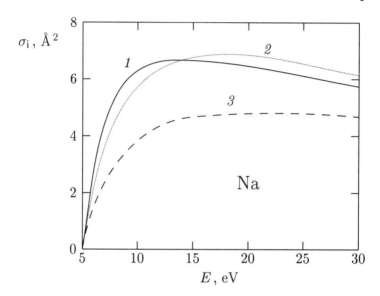

Fig. 28. The electronic collision ionization cross section for the sodium atom as a function of the incident particle energy: *1*— experiment [113]; *2* — calculation by the Born – Compton method; *3* — calculation according to Eletsky – Smirnov's formula [108].

The results of calculations of the ionization cross section in electronic collision with neutral atoms within the frames of the discussed approaches, as well as the corresponding experimental data, are shown in Fig. 28 – 31. Note that for the hydrogen atom, the BEB approximation [109] (see Eqn. (9.19)) yields the best agreement with experimental data [112]. The BC approach (in both modifications) overestimates appreciably the real cross section and shifts the maximum toward low energies compared to the experimental curve. This situation is typical for targets with a high ionization potential ($I_p > 10$ eV), as well as for light atoms and ions. Figure 28 shows the collisional ionization cross section for an alkali atom (sodium) ($I_p = 5.139$ eV). Clearly, the BC (a) method better fits experimental data [113] than simple approaches known earlier.

The relevant information is also collected in Table 4. Here E_m is the IP energy at the cross section maximum. It is clear from

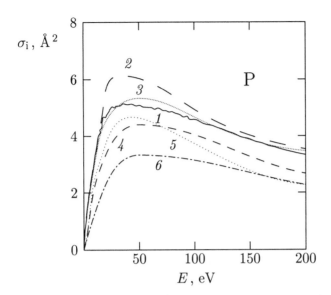

Fig. 29. The electronic collision ionization cross section for the phosphorus atom as a function of the incident particle energy: *1* — experiment [110]; *2* — calculation by the Born – Compton method (a); *3* — calculation by the Born – Compton method (b); *4* — calculation according to Eletsky – Smirnov's formula [108]; *5* — calculation using Lotz's formula [107]; *6* — calculation by the BEB method (the simplest modification) [109].

Table 4.

	LiI	NaI	KI	RbI	CsI
E_m, eV, exp [113]	13	14	8.5	10.5	9.5
E_m, eV, BC(a)	17.8	17	14.32	13.8	12.84
E_m, eV, ES	21.6	20.6	17.36	16.7	15.56
σ_{imax}, Å^2 [113]	4.2	6.8	7.9	-	10.2
σ_{imax}, Å^2 [114]	—	8.6	9.6	9.6	11
σ_{imax}, Å^2 [115]	4.9	7.6	8.2	8.2	9.4
σ_{imax}, Å^2 BC(a)	6.2	6.8	9.4	10.3	11.9
σ_{imax}, Å^2 ES	3.39	4.9	6.9	7.45	8.6

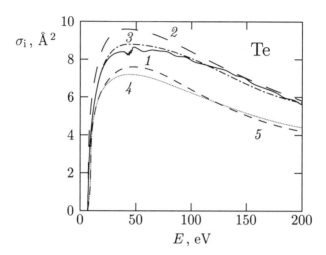

Fig. 30. The electronic collision ionization cross section for the tellurium atom as a function of the incident particle energy: *1* — experiment [110]; *2* — calculation by the Born – Compton method (a); *3* — calculation by the Born – Compton method (b); *4* — calculation according to Eletsky – Smirnov's formula [108]; *5* — calculation using Gryzinski's formula [106].

this Table that the experimental values E_m for the alkali atoms are better reproduced by the BC(a) method than by Eletskij – Smirnov's semi-empirical relation (9.18). This statement is also valid for the maximum value of the ionization cross section σ_{imax} with the exception of the lithium atom. As follows from the shape of the curves in Fig. 27, the biparametric form (9.19) of the BEB approximation [109] for the collisional ionization of the alkali metal atoms yields results in worse agreement with experiment than the BC method.

The asymptotic behavior of the collisional ionization cross section for the alkali atoms inside the energy range from 30 to 500 eV was experimentally reported in paper [114]. This dependence proves to be the same for all alkali atoms and is described by the following formula [114]:

$$\sigma_i = \alpha \, E^{-\beta} \, . \tag{9.21}$$

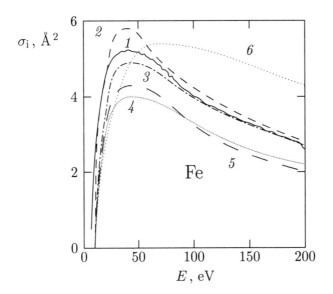

Fig. 31. The electronic collision ionization cross section for the iron atom as a function of the incident particle energy: *1*— experiment [110]; *2*— calculation by the Born – Compton method (a); *3*— calculation by the Born – Compton method (b); *4*— calculation using Eletsky – Smirnov's formula [108]; *5*— calculation using Gryzinski's formula [106]; *6*— calculation by the BEB method (the simplest modification) [109]. Calculations of plots *2 – 5* were carried out ignoring contributions from *d*-electrons.

The best fit to the experimental data is provided by choosing $\beta = 0.592$. This value of the parameter is close to the BC model prediction $\beta = 0.5$ (see (9.20)) and differs significantly from values given by other approaches, as is clear from Eqns (9.17) – (9.19).

Calculated and experimental data [110] for collisional ionization of atoms with filling electronic np-subshells ($n = 3, 5$) in the IP energy range from the ionization threshold to 200 eV are presented in Fig. 29 – 30.

The above analysis suggests that the BC model is more consistent with experiments than the known simple approximations for atoms with moderate ionization potentials from groups III – VI (with the exception of the aluminum and sulphur atoms) of the Periodic Table. Note that light atoms of the II period provide exceptions as well. The collisional ionization cross section for those atoms are anomalously small from the viewpoint of the BC model predictions.

The phosphorus and tellurium atoms show particularly good agreement. Calculations within the BEB approximation [109] using Eqn. (9.19) are given only for phosphorus atoms. For realistic values of the reduced kinetic energy of an electronic subshell, the BEB approach yields a cross section smaller than given by Eletskij – Smirnov's semi-empirical formula which, as the presented plots indicate, underestimates somewhat the actual cross section in the considered cases. Note that a more accurate modification of the BEB approximation would require more detailed information, namely, the differential oscillator strength of electrons from the subshell being ionized. Essentially it drops out from the class of simple methods for collisional ionization cross section calculations.

Calculations show that for the incident electron energies below the value corresponding to the cross section maximum, the best results are obtained as a rule by the (a)-modification of the method suggested (Eqn. (9.20a)), evidenced especially by the silica and antimony atoms. For large incident particle energies the (b)-modification of the BC approximation is preferred (Eqn. (9.20b)). Moreover, the (a)-modification better fits experimental data for atoms from the first half of the period, while the (b)-modification is more appropriate for the 2d half.

Figure 31 demonstrates results for the ionization cross section of the iron atoms by electronic impact. The calculations involving single-parametric similarity functions (9.17), (9.18), and (9.20) have been done ignoring the contribution from d-electrons to the total cross section which has been assumed to be negligible. Note that the BEB method accounts for the ionization cross section decrease for electronic subshells with large values of the reduced kinetic energy (it corresponding to a large orbital angular momentum). Nevertheless, appreciably overestimates the contribution from d-electrons to the ionization cross section for the iron group atoms. Figure 31 implies that the collisional ionization cross section for the iron atoms measured in experiment lies between the corresponding curves for (a) and (b) modifications of the Born – Compton approximation. The calculation shows that in the case of the nickel atom better results are obtained by the (b)-modification of the BC method. It is interesting to note that the best approximation to the experimental cross section for the collisional ionization cross section of the copper atom is obtained by choosing the effective number of $4s$ electrons to be $N_{4s} = 1.5$, which correlating with the peculiarity of filling electronic subshells in passing from the nickel to copper atom.

To summarize, we note that in this Section, using model considerations, a new simple method for calculation of collisional ionization of neutral atoms is developed, the Born – Compton approximation, in which the dynamics of atomic electrons is described using the Compton profile of X-ray scattering by target's electrons. Using universal approximation of the Compton profile (9.3) allows the collisional ionization cross section to be expressed through the corresponding similarity function, representing a double integral of an elementary function and depending only on the incident particle energy ratio to the ionization potential of the ionized subshell. Two modifications are suggested depending on the choice of the upper limit of energy transmitted to the displaced electron, each of them with its own preferable applicability region. Simple approximate expressions are found for the Born – Compton similarity functions that allow rapid cross section calculations.

Comparison with existing experimental data and known simple calculation methods indicates that the approach proposed seems to be preferential for neutral atoms "from the middle" of Mendeleev's Periodic Table with moderate ionization potentials, such as elements from 3 – 5th periods of groups III – VI. In addition the proposed method satisfactorily describes the collisional ionization cross section for the iron group atoms ignoring the d-electron contribution to the process. An important advantage of this method is its simplicity: the minimum number of atomic parameters is required to calculate the cross section. At the same time the new method overestimates the process cross section for atoms with high ionization potential and light atoms from the second period of the Periodic Table, by shifting the maximum toward low energies, so the employment of traditional methods, such as the semi-empirical Eletskij – Smirnov's formula [108] or Kim's BEB approximation [109] appears to be preferrable.

Thus the Born–Compton approximation considered here can be used as a simple method for calculating collisional ionization cross sections of those atoms for which traditional methods known earlier yield unsatisfactory results.

9.3. Polarization bremsstrahlung radiation with ionization of atom: relation with X-ray Compton scattering

Polarization bremsstrahlung radiation (PBR) of a charged particle on target with an electronic core is connected with conversion of the incident particle proper field into a real photon on the bound electrons of the core [9]. Depending on the momentum q transmitted to the target during the radiative process, PBR can be coherent ($q < p_a$, p_a is the characteristic momentum of bound electrons) or incoherent ($q > p_a$) with respect to the contribution of individual bound electrons to the total radiation. In the last case PBR is accompanied, as a rule, by ionization of the atom [116], so it has the second name, "radiative ionization" [117]. Note that PBR can be accompanied by the excitation of an atom, too [116].

Energy conservation implies that PBR is always incoherent within the frequency range $\omega > p_a v_0$ (v_0 is the incident particle velocity).

Incoherent PBR, whose cross section is proportional to the number of the atomic electrons [116], can be interpreted as was noted above, as being the Compton scattering of the incident particle proper filed on target's electronic core, during which some fraction of the energy transmitted ionizes the atom. Such an approach has been tightly related to the equivalent photon method of E. Fermi [22], who considered the interaction of a charged particle with an atom as its irradiation by the electromagnetic pulse of the equivalent photons. This approach allows using the known photo ionization cross sections to calculate collisional ionization of atoms by fast electrons. In this case the establishing of a relationship between PBR and the Compton scattering of photons is important for PBR studies, since there is extensive information on the Compton scattering for all elements from the Periodic System, (see for example [105]).

Under experimental conditions the process of incoherent PBR of a non-relativistic electron on an multi-electron atom is as a rule is fully masked by the bremsstrahlung radiation of the secondary electrons via the ordinary static channel. Incoherent PBR may appear either for a relativistic particle (due to different angular distribution of the bremsstrahlung photons in polarization and static channels), or for a heavy incident particle, when the static channel is suppressed. So incoherent PBR in the non-relativistic case appears to be physically interesting mainly for scattering of heavy particles.

Incoherent PBR is studied below for multi-electron atom – targets that allow a universal description of the process cross section, valid for all elements from the Periodic Table, obtainable using the analogy between PBR and X-ray Compton scattering on atoms.

At high frequencies $\omega \gg I$, the PBR spectral cross section for a non-relativistic Born charged particle, integrated over the photon emission solid angle, reads in ordinary (Gaussian) units:

$$d\sigma(\omega) = \frac{8}{3\pi} \frac{e^4 e_0^2}{m_e^2 v_0 \hbar c^3} \frac{d\omega}{\omega} \iint d\Omega_{\mathbf{q}} \, dq \, S(q^0, \mathbf{q}) \qquad (9.22)$$

where $e_0 = Z_{\mathrm{pr}} e$ is the IP charge.

Note that within the approximation of quasi-free (rest) atomic electrons, the *incoherent* DFF of the target has the form:

$$S_{\mathrm{free}}^{\mathrm{ncoh}}(q) = \frac{Z}{q \, v_0} \delta\left(\frac{\omega + \mathbf{q}\mathbf{v}_0 + \mathbf{q}^2/(2\,\mu)}{q \, v_0}\right) \qquad (9.23)$$

Here μ is the reduced mass of the electron and IP, Z is the number of atomic electrons which is equal to the nuclear charge.

Considering the DFF relation with Compton profile $J(Q)$ (9.8) and the CP scaling (9.3), the following representation of the spectral PBR cross section at frequencies $\omega > p_a v_0$ is obtained from Eqn. (9.22):

$$d\sigma(\omega, v_0, m) = \sqrt[3]{Z} \, Z_{\mathrm{pr}}^2 \, d\tilde{\sigma}(\omega \, r_{\mathrm{TF}}^2, v_0 \, r_{\mathrm{TF}}, m)\,. \qquad (9.24)$$

Here we introduced the reduced cross section $d\tilde{\sigma}$ depending upon the emitted photon frequency and the IP velocity, appropriately normalized to the characteristic radius of the Thomas – Fermi atom and denoted below by variables with the tilde. The reduced cross section is expressed through the normalized Compton profile of the atom

$$d\tilde{\sigma}(\tilde{\omega}, \tilde{v}, m) = \sigma_0 \frac{b^2}{\tilde{v}^2} \frac{d\tilde{\omega}}{\tilde{\omega}} I(\tilde{\omega}, \tilde{v}, m), \qquad (9.25)$$

$$I(\tilde{\omega}, \tilde{v}, m) = \int\limits_{q_{\min}}^{q_{\max}} \frac{d\tilde{q}}{\tilde{q}} \int\limits_{-v}^{-v_m} \left\{ \tilde{J}\left(-\tilde{q} + \sqrt{-2\,\tilde{q}^0}\right) \right.$$

$$\left. - \tilde{J}\left(\tilde{q} + \sqrt{-2\,\tilde{q}^0}\right) \right\} d(v_0 \cos(\mathbf{q}\,\mathbf{v}_0)). \qquad (9.26)$$

Here $v_m = \left(\tilde{\omega} + \tilde{q}^2/2\,m\right)/\tilde{q}$, $b = 0.8853$. The upper and low integration limits on the modulus of the transmitted momentum in integral (9.26) are determined by the condition $v_m < v$: $q_{\min, \max} = m_0 v_0 \left[1 \mp \sqrt{1 - 2\,\omega/(m_0\,v_0^2)}\right]$. The dimensional cross section σ_0 entering expression (9.25) is

$$\sigma_0 = \frac{16}{3} \frac{e^2}{\hbar\,c} \left(\frac{\hbar}{m_e\,c}\right)^2 = 2.074 \cdot 10^{-6} \text{ a.u.} \qquad (9.27)$$

Thus Eqn. (9.25) and (9.26) reveal the similarity law for the incoherent PBR cross section of a fast (but non-relativistic) charged particle on multielectron atom and express the process cross section through the normalized X-ray Compton scattering profile. This cross section (within the factor $\sqrt[3]{Z}$) depends on the emitted photon frequency and the IP velocity expressed in units of the Thomas – Fermi momentum.

Note that although similarity law (9.25) – (9.27) has been obtained, strictly speaking, within the statistical model frames, it is approximately valid (to the "smearing out" of the shell structure) for the Hartree – Fock atom as well due to the aforementioned approximate scaling for the normalized Compton profiles (see Fig. 26).

For comparison, we present the corresponding expression for the incoherent PBR cross section on a hydrogen-like ion with charge Z:

$$d\sigma_H(\omega, v, m) = Z^{-1} b^{-2} d\tilde{\sigma} \left(\frac{\omega}{p_H^2}, \frac{v}{p_H}, m \right) \qquad (9.28)$$

where $p_H = Z$ a.u.

. Equation (9.25) for the high frequency PBR cross section with ionization of atom specifies and completes the result [116] obtained using DFF in the model of free atomic electrons (9.23). This can also be represented in form (9.25) and (9.26) provided that

$$I_{\text{free}}(\tilde{\omega}, \tilde{v}, \mu) = \ln \left\{ \frac{1 + \sqrt{1 - 2\tilde{\omega}/(\mu\tilde{v}^2)}}{1 - \sqrt{1 - 2\tilde{\omega}/(\mu\tilde{v}^2)}} \right\}. \qquad (9.29)$$

A characteristic feature of the process is the appearance of the cut-off frequency caused by momentum conservation. From kinematical considerations the "cutoff" frequency for electron PBR is two times smaller than for proton PBR (due to the difference in the reduced masses). This justifies the above conclusion for the independence of the PBR cross section of the incident particle mass.

Spectral dependences of the effective incoherent proton PBR $\omega \frac{d\sigma}{d\omega}$, calculated using different approximations, including the model of free atomic electrons, are shown in Fig. 32. Clearly the main difference between the models appears at frequencies $\omega > \omega^* = \mu v^2/2$; i.e., above the "cutoff" frequency for PBR on free electrons.

The cross section decrease (with increasing PBR frequency in the exponential screening model) occurs much faster than for the Hartree – Fock Compton profile, also evident from Fig. 26. At frequencies below the "cutoff" frequency, the Hartree – Fock method of accounting for coupling of atomic electrons in the initial state yields a somewhat smaller cross section compared to the model of free atomic electrons.

Note a close similarity of dependences shown in Fig. 32 to the corresponding spectral cross sections for radiative ionization from paper [117]. In this paper a similar approach was applied to describe incoherent PBR (radiative ionization) based on using the non-diagonal atomic form factor $F_{n,W}(q)$, calculated earlier in connection with

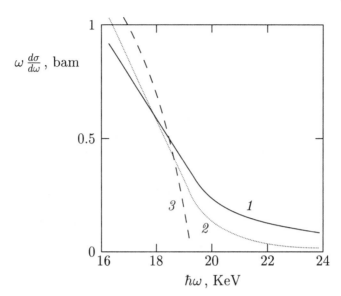

Fig. 32. The spectral cross section of non-coherent PBR of a proton with energy 34 MeV on the krypton atom near the 'cut-off' frequency calculated within the frames of different approximations for the atomic electron density: curve *1* — the Hartree–Fock calculations. curve *2* — the exponential screening, curve *3* — the approximation of free atomic electrons.

the atomic ionization problem [118] and the emission of characteristic X-rays.

Figure 33 show the effective emission $\omega \frac{d\sigma}{d\omega}$ of the incoherent proton PBR on the krypton atom for three values of the bremsstrahlung photon energies 3.78, 7.57 and 11.35 keV. It is seen that these functions have maxima that shift toward higher velocities with increasing the bremsstrahlung photon energy. The corresponding formula relating the bremsstrahlung photon frequency with the optimal value of the proton velocity has the form (in atomic units):

$$v_{\mathrm{opt}} = 1.89 \sqrt{\omega}. \tag{9.30}$$

It is essential that relationship (9.30) is independent of the nuclear charge, opposite to the similar dependence for coherent PBR, when the optimal velocity linearly depends on the emitting frequency [119] through the atomic subshell radius that mostly contributes to the process.

Note that the incoherent PBR cross section (9.24) is proportional to the square of the incident particle charge. Its intensity can be very significant for heavy multi-charged ions that are used in modern experiments on storage rings (see for example [66]).

Now we wish to also analyze the relation between the coherent and non-coherent PBR cross sections. The cross section for the coherent process will be calculated within the exponential screening model for the electron density of target's core. The corresponding expression can be derived from general formula (9.22), bearing in mind that the atomic DFF in this case is reduced to the ordinary static form factor, a Fourier-image of the electron density.

After standard transformations including integration over the emitted photon solid angles and transmitted momenta, the coherent

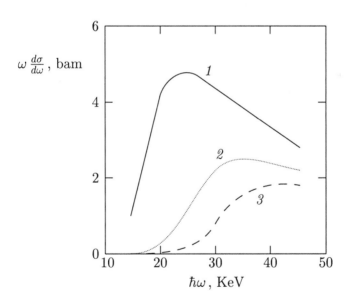

Fig. 33. The non-coherent proton PBR on the krypton atom as a function of the proton velocity for three values of the bremsstrahlung photon energy: $\hbar\omega = 3.78$ keV (curve 1), $\hbar\omega = 7.57$ keV (curve 2), $\hbar\omega = 11.35$ keV (curve 3).

PBR cross section of a non-relativistic Born particle is found to be

$$d\sigma_{\text{coh}}^{(\text{exp})}(\omega) = \frac{16}{3} \frac{Z^{4/3} Z_0^2}{\tilde{v}^2 c^3} b^2 \int\limits_{\tilde{q}_{\min}}^{\tilde{q}_{\max}} \frac{1}{\left(1 + \tilde{q}^2/2\right)^4} \frac{d\tilde{q}}{\tilde{q}} \frac{d\omega}{\omega}. \qquad (9.31)$$

In the integral from Eqn. (9.31) are the same limits for the transmitted momentum as in Eqn. (9.26). The "tilde" above the transmitted momentum sign and the IP velocity again indicates the normalization to the momentum (velocity) of the Thomas–Fermi atom.

The integral in Eqn. (9.31) can be taken in quadratures, but the resulting expression is sufficiently awkward. For a heavy IP, the upper integration limit can be substituted by infinity. Then the integral over the transmitted momentum becomes

$$I_{\text{coh}}^{(\text{exp})}(\tilde{\omega}, \tilde{v}) = \frac{11 + 54\left(\frac{\tilde{v}}{\tilde{\omega}}\right)^2 + 72\left(\frac{\tilde{v}}{\tilde{\omega}}\right)^4}{12\left(1 + 2\left(\frac{\tilde{v}}{\tilde{\omega}}\right)^2\right)^3} - \frac{11}{12} + \frac{1}{2}\ln\left(1 + 2\left(\frac{\tilde{v}}{\tilde{\omega}}\right)^2\right).$$
$$(9.32)$$

Note that in the high frequency limit $v/r_{\text{TF}} \ll \omega$ formula (9.32) has the asymptotic

$$I_{\text{coh}}^{(\text{exp})}(\tilde{\omega}, \tilde{v}) \cong 2 \, (\tilde{v}/\tilde{\omega})^8. \qquad (9.33)$$

The inequality $v/r_{\text{TF}} \ll \omega$ can be recasted in the form: $\omega \gg 0.125 \, Z^{2/3}$ KeV, implying that it is valid for all Z the keV energies of the bremsstrahlung photon.

Collecting together all above formulas, we arrive at the coherent PBR cross section in the exponential screening approximation of the high frequency limit $\omega \gg 0.125 \, Z^{2/3}$ KeV:

$$d\sigma_{\text{coh}}^{(\text{exp})}(\omega) = \frac{32 \, b^2}{3 \, c^3} Z^{4/3} \frac{\tilde{v}^6}{\tilde{\omega}^8} \frac{d\omega}{\omega}. \qquad (9.34)$$

For correct assessment of the relation between the coherent and incoherent cross sections of the process it is important to stress that simple approximation of the atomic density by one exponent with the characteristic cut-off at the Thomas–Fermi radius underestimates significantly the K-shell contribution to coherent PBR on multielectron atom at high frequencies. Indeed, the closest orbit to the nucleus has a radius about $Z^{2/3}$ times smaller than the Thomas – Fermi radius. Thus the corresponding integral in Eqn. (9.31) leads to another form of drop-off in the spectral cross section high-frequency than for the Thomas – Fermi radius.

Taking into account an additional contribution from the K-shell, we rewrite formula (9.31) in the form ($Z \gg 1$):

$$d\sigma_{\text{coh}}^{(\text{exp})}(\omega) = \frac{16}{3} \frac{Z^{4/3}}{\tilde{v}^2 c^3} b^2 \frac{d\omega}{\omega}$$
$$\times \left\{ I_{\text{coh}}^{(\text{exp})}(\tilde{\omega}, \tilde{v}) + \frac{4}{Z^2} I_{\text{coh}}^{(\text{exp})}(\tilde{\omega}, \tilde{v} (p_K(Z)/p_{\text{TF}})) \right\}. \quad (9.35)$$

Here we have introduced the K-shell momentum $p_K(Z)$.

Equation (9.35) is a universal representation (for all nuclear charges) of the coherent PBR cross section of a fast particle obtained within the framework of the exponential electron density model, with a separate account of the K-shell contribution to radiation.

Coherent and incoherent PBR cross sections of proton on the krypton atom for two proton velocities are shown in Fig. 34. This figure implies, in particular, that the incoherent process can prevail over the coherent one for sufficiently high incident particle velocities. Indeed, in that case the "cut-off" frequency of the radiative ionization shifts toward higher frequencies where the contribution of most atomic electrons to coherent PBR becomes small.

In this Section we have found a universal description of incoherent PBR of fast charged particles on multielectron atoms. In PBR calculations, a relationship between the process with the Compton X-ray scattering profile has been established. This allows extensive data for Compton scattering to be used in the PBR calculations. The universal description is based on an approximate scaling of the reduced X-ray scattering Compton profile on neutral atoms which is valid for sufficiently large nuclear charges ($Z \geq 20$). Based on formulas obtained, we have analyzed spectral and velocity dependences of heavy charged particle incoherent PBR on multielectron atom. For a given PBR frequency, an optimal incident particle velocity has been shown to exist at which the process cross section reaches maximum.

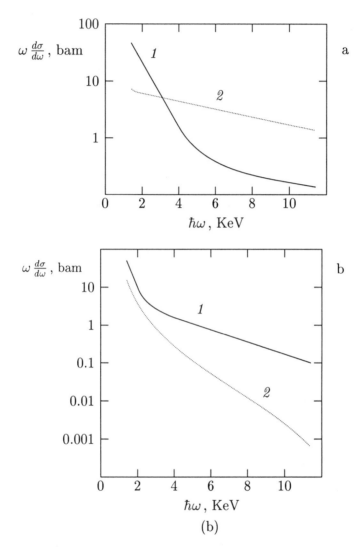

Fig. 34. Spectral cross section of the non-coherent (curves *1*) and coherent (curves *2*) proton PBR on the krypton atom for different proton velocities: (a) $Z=36$, $v = 10r_{\mathrm{TF}}^{-1} = 37.3$ a.u., (b) $Z=36$, $v = 3r_{\mathrm{TF}}^{-1} = 11.2$ a.u.

The optimal velocity increases as the square root of the frequency. The comparison of the coherent and incoherent cross sections has discovered that at sufficiently high incident particle velocities there exists a frequency range inside which PBR with an atom's ionization dominates over coherent PBR. Experimental aspects are discussed in Sect. 10.

10. Experimental aspects

Polarization effects in atomic transitions are studied experimentally to a lesser degree than theoretically. One of the first experimental applications of the PBR theory was bringing into agreement experimental data on the laser break-down of the alkali metal vapors [99]. This included predictions of the avalanche theory using the inverse bremsstrahlung effect cross section calculated in [91] with account for the polarization mechanism.

In this section, we consider experiments on polarization BR and related phenomena for different media and energies of incident particles.

10.1. Experiments on PBR of electrons on atoms

Polarization BR in scattering of the keV-electrons on isolated atoms was first observed in paper [120], in which a supersonic Xe stream was used as a dense atomic target. Measurements were carried out in the spectral range corresponding to the "giant" resonance in the photoabsorption cross section caused by the $4d$-subshell (the maximum frequency $\omega_{max}^{ph} \approx 98$ eV, the line half-width $\Gamma \approx 24$ eV). The contribution due to polarization channel is most significant in this region [121]. The emission spectrum form was found to be very similar to that of the photoabsorption line. This is explained by the relation of the imaginary part of the dynamic polarizability, which determines the PBR cross section in this case, with the photoabsorption cross section, consistening with the optical theorem. However, for xenon, as the estimations from paper [121] suggest, the ratio of the PBR cross section to that of static BR is less than unity even at the maximum (around 0.4). In order to distinguish the contribution due to the polarization channel a special procedure was employed in [120] based on subtracting the argon emission spectrum from the observed one and for which the PBR contribution in the spectral range under consideration is negligible. Experimental conditions were chosen such that for photons outside the "giant" resonance in the $4d$-absorption, the intensities of emission spectrum of argon and xenon coincided. It is essential that the spectral range under study (70 – 150 eV) does not contain the line spectrum from the considered species. The energy of electron beam was 600 eV with the current density 0.3 A/cm^2. The electron energies were bounded from above since ionization of the $3d$-subshell of xenon is possible at high energies. Based on these measurements the cross section ratio of PBR and static BR was shown to be about 70%. Maximum of the emission observed was at a photon energy of 113 eV, which significantly exceeds the ionization threshold of the $4d$-subshell of xenon and is shifted towards high energies in comparison with the photoabsorption maximum at about 20 eV.

Emission spectra of oxides and fluorides of lanthanum and thorium caused by the bombarding of solid targets by keV-energy electrons were studied in papers [122, 123]. The ratio of BR channels at a frequency near the ionization potential of the $4d$-subshell of lanthanum is (according to the estimate made in [121]) around 3.2. Thus it is lanthanum that is of the uppermost interest from the point of view of observations of polarization effects in BR. At frequencies above the ionization threshold of the $4d$-subshell of lanthanum ($110\,eV < \omega < 140\,eV$) studied in [122, 123], where the "giant" resonance occurs, photoabsorption by metallic lanthanum is very close to the atomic one. This provided the grounds to consider the emission spectrum to be of atomic character. The emission studied was excited by electrons with energies from 500 eV to 4 keV with a current strength of 4 – 6 mA. The energy resolution of the photodetector used VEU-6 was 0.18 eV at a photon energy of about 100 eV. The samples examined were taken in the form of thin films around 100 Angstroms in thickness, which allowed avoidance of the radiation self-absorption that mimics the measured spectrum in a volume sample, especially at high incident particle energies [123].

The measured emission line profile was found to be very close to that of the photoabsorption line. The position of the emission maximum was shifted by about 0.5 eV toward low energies compared to the absorption maximum, and the spectral width of the band in emission was somewhat larger than in absorption. As noted in paper [123], the central frequencies of the spectral emission maximum for oxides and ftorides of lanthanium coincide. Moreover, it was discovered that the corresponding line half-width in the LaF_3 sample (4.1 eV) is smaller than in the La_2O_3 sample (5.7 eV). Thus, the registered spectra proved to be sensitive to some degree to the chemical composition of the sample.

It is quite symptomatic that the emission line profile was asymmetric with a slower decrease at the high frequency wing. This fact can be explained by cross-channel interference, known [93] to have a destructive character in the low-frequency wing and a constructive character in the high-frequency wing.

As was noted in paper [121], the Born approximation is, strictly speaking, insufficient to describe the emission spectrum obtained in experiments [107] using the 0.5-keV electrons. However it was used to describe qualitative relationships. The spectrum calculated in [121] using a formula similar to Eqn. (5.3) turned out to be very close to that measured in [123]; here at all frequencies the contribution from the polarization channel was larger than that from the static one.

Experimental studies of PBR of the keV-energy electrons on the xenon atom near the ionization threshold of the $4d$-subshell were continued in papers [124, 125]. Paper [124], in addition to reporting on experimental results, suggests a theoretical explanation for the discovered emission spectrum maximum shift with the change in the IP energy and the bremsstrahlung photon's emission angle. Increasing the electron energy from 0.3 keV to 0.9 keV resulted in the maximum spectrum energy shift from 123 eV to 105 eV with a photon emission angle of 97°. Here the line profile asymmetry was found to strengthen such that the intensity shortly damped toward high frequencies. Moreover, for electron energies 0.3 keV and 0.6 keV, additional maxima in the high-frequency emission line wing appeared. The explanation to the observed shift of the maximum was based on calculating the differential over the photon angle PBR cross section, in which an additional term was accounted for in addition to the logarithmic one. A simple expression was obtained for the frequency detuning of the emission spectrum maximum off the photoabsorption maximum, predicting this parameter to decrease with the IP energy, an event actually observed in experiments. The computed values of the detuning in the high-energy IP limit were found, however, to exceed the experimental ones. This can apparently be explained by the inadequacy at the Born approximation.

Measurements of the differential BR spectral intensities in relative and absolute units for different energies of electrons scattering on the xenon atoms were carried out in paper [125]. The dependence of the PBR intensity on the IP energy turned out to have a maximum lying inside the 0.6 – 0.7 keV electron energy range, contradicting the Born approximation predictions; in this case, however, its applicability is not justified. A dependence different from Born's was also observed for the BR intensity as a function of the IP energy in the static channel. By increasing the electron energy from 0.3 to 0.6 keV, a linear growth of intensity was registered, notably slowing down in the 0.6 – 0.9 keV range. At the same time, as is well known [9, 126], the Born approximation yields the inversely proportional dependence of the BR spectral intensity on the IP energy for both channels. The deviation from the Born relationship confirmed in [125] is mainly due to effects of the IP penetration into the target's electronic core (see Sect. 5.3 above).

10.2.   Relativistic experiments on accelerators

Experimental studies of polarization effects in relativistic electron radiation in condensed media [127 – 130] have been intensively developed recently.

Paper [127] reported on observations of polarization-interferometric features in the emission of relativistic electrons (with energies of 15 MeV and 25 MeV) in a silicium crystal 50 mcm in thickness. In that paper the interference of the coherent and parametric mechanisms for the X-ray emission was first discovered. The parametric X-ray radiation is synonymous with the coherent PBR that had been used for describing relativistic electron emissions in periodic media long before the general concept of PBR had appeared. In order to observe the effect, a crystallographic plane (220) was used in this experiment which proved to be most convenient under the given experimental conditions. In the course of coherent radiation the momentum excess equal to the inverse crystal lattice vector corresponding to the plane (220) was transmitted to the medium. If the transferred momentum is specified then there is a notably increase in the cross-channel interference compared to the radiation in an amorphous medium. In the last case the summation over possible transferred momenta "smears out" the interference effect. The photon emission angle was fixed by the registration channel of the setup to be 0.31 rad. The orientation dependence of the photon production into a solid angle of $\Delta\Omega = 1.23$ sr on the angle between the electron beam axis and the crystallographic plane (220) was measured in the experiment. In the case under study the coherent PBR is the main contributor. The cross-channel interference, constructive for some angles and destructive for other angles, alters only slightly the total output photon intensity. This change was reliably fixed by the experiment, the interference effect being more pronounced for smaller electron energies (15 MeV). The results of measurements proved in good agreement with calculations [131].

The PBR suppression effect for a relativistic electron moving through a thin amorphous carbon foil was reported in paper [128]. A continuous electronic beam with energy $5-7$ MeV was employed. The current in the beam was a few nano-Amperes. Photons with energies $2-30$ keV were registered by a semiconductor Si(Li) cooled detector. The observations were carried out under an angle of 45 degrees to the electronic beam propagation direction. A more rapid decrease of the PBR intensity with frequency was detected than that predicted by theory of the process on isolated atoms. The explanation to the effect suggested in this paper is based on the effective screening radius of the nucleus by the atomic electrons increasing in a condensed medium. This results in the decrease of the maximum frequency $\omega_{\rm m} = v_{\rm i}/R_0$ above which the coherent PBR cross section is effectively suppressed. Indeed, in the case under study four out of six electrons of the silicium atom participate in establishing chemical links with neighboring atoms, so that the effective atomic radius $R_0^{\rm eff}$ increases. Calculation involving the effective radius $R_0^{\rm eff} = 4.5R_0$

(R_0 is the radius of an isolated silicium atom) proved to be consistent with experimental data. In the case under consideration the density effect in PBR, due to the destructive interference of contributions from neighbor atoms (see Sect. 6.3), operates in a much more longwave range, inaccessible for registration by the detectors used in the experiment. In [128] the contribution of the non-coherent PBR (with the target's excitation and/or ionization) was also evaluated. The experimental data were found to approach an intensity level corresponding to the non-coherent PBR for photon energies above 5 keV. Unfortunately, in this paper there was no possibility to observe photons with energies below 3 keV, so the PBR suppression effect was detected in a narrow energy range between 3 keV and 5 keV. Nevertheless, the results obtained in [128] do suggest that PBR can be used as a tool for structural studies of the medium.

In the recent paper [129] (see also [130]), the relativistic electron PBR in the polycrystal aluminium was examined. The spectral-angular distribution of the PBR intensity was measured for electrons with an energy of 2.4 MeV in the photon energy range from 2 keV to 8 keV in the interaction of the electronic beam with a thin aluminium foil. The radiation was detected with a Si(Li) detector in a small solid angle ($\approx 1.5 \times 10^{-3}$ sr) under the angle 90 degrees with respect to the electron beam propagation direction. The target's surface plane was set under the 45 degrees to the beam axis. The distance between the target and the detector was about 0.5 m. Note that in this case the main contribution to the total number of the registered photons is given by the K-peak of the characteristic radiation of aluminium near the energy 1.5 keV. According to theory, the PBR intensity under the given experimental conditions is not more than 1% of this characteristic radiation intensity which, of course, was taken into account in processing the experimental data. Also, in order to obtain the PBR spectrum the background radiation should be also subtracted. It was found that the PBR spectrum of relativistic electrons in the polycrystal aluminium has sharp peaks due to the coherent scattering of the proper IP field on the ordered structure of microcrystals forming the polycrystal target. This is in a sharp contrast with the frequency dependence of PBR in an amorphous medium, where, according to a calculation made for experimental conditions [129], the PBR intensity monotonically decreases with photon energy. The nature of the PBR spectral maxima discovered is similar to the Debye – Shearer peaks in the X-ray scattering in polycrystals [132]. Results [129] also suggest that the PBR intensity outside the coherent peaks is strongly suppressed compared to the emission in an amorphous medium. The theoretical curve for the PBR spectral intensity proved to be in good correspondence with experimental data [129]. Note that the intensity of the ordinary (static) BR in the spectral-angular range under study was by about 4 times smaller than the BR intensity registered in the vicinity of the principal spectral peak at the photon energy 4 keV. Thus, in [129], the frequency dependence for the relativistic electron PBR in a polycrystal medium was found to be strongly different from the case of an amorphous target [127], unlike the static BR. PBR turned out to be very sensitive to the target structure, which signals good prospects for the elaboration of novel diagnostic methods in solid bodies by using this phenomenon.

10.3. Polarization bremsstrahlung of heavy charged particles

An important characteristic property of the polarization BR is a weak dependence of its cross section on the IP mass [9]. In the static BR case, as is well known [44], the situation is radically different. Here the process cross section is inversely proportional to the square of the reduced mass of the colliding particles, so the static BR of heavy charged particles is negligible compared to the static BR of electrons and positrons.

The total proton BR on the hydrogen atom with account for the polarization channel was first calculated in paper [133]. In this paper, the BR cross sections for protons and electrons at frequencies $\omega \leq \omega_m = v_i^2/2$ have the same order of magnitude. At frequencies $\omega < v_i/R_0$ the process proceeds without target excitation ("elastic" or coherent BR); if $\omega > v_i/R_0$, BR is accompanied by the atom's ionization — this is "inelastic" or non-coherent BR (see Sect. 9.3).

The total BR cross section generated by light ions in collision with multielectron atoms was calculated in [117]. This work was stimulated by experiments carried out by the same authors on BR of protons and ^{3}He ions in the scattering on a thin aluminium foil [134]. In [134] a significant discrepancy was discovered between experimental data on radiation from protons with energies 1 MeV and 2 MeV in the photon energy ranges $2-6$ keV and $4-6$ keV, respectively, with theory predictions that ignored the polarization channel. At the same time, the experiments revealed good correspondence with the theory for emission of protons with energies above 3 MeV in the entire frequency range. The calculation performed in [134] included two radiative mechanisms: static BR of a heavy charged particle and BR of secondary electrons knocked out by this particle from target's atoms. Note that in the case of multicharged heavy IP, dominating radiative mechanisms can also include a radiative target's electron capture into the states from the IP discrete spectrum [135] and X-ray emission of molecular orbitals [136]. However, for light ions the relative contribution of these processes to the total emission is small. Calculations [134] revealed that the secondary electron BR cross section in the given experimental conditions is larger by $3-4$ orders of magnitude than the static proton BR cross section, so the latter radiative mechanism can be disregarded in analyzing the experimental data. A characteristic feature of the IP and bremsstrahlung photon energies studied in [134] is their closeness to the secondary electron BR "cut-off" frequency (i.e. to the frequency above which the process is prohibited by conservation laws), which is $\omega^{\mathrm{h}} = 2v_{\mathrm{i}}^2$, i.e. grows linearly with the IP energy. The concept of the polarization (or atomic) BR was drawn in [117] to explain the above mentioned discrepancy between theory and experiment. In this paper both coherent PBR (without target's excitation or ionization) and non-coherent PBR (radiative ionization in terms of paper [117]) were computed. The contribution due to the latter process proved insignificant for the interpretation of experiments [134]. The point is that in the low-frequency part of the considered range the coherent PBR cross section, which is proportional to the square of the number of atomic electrons N_{a}, exceeds that of the non-coherent process which is linear on N_{a}. In the high-frequency part of the bremsstrahlung spectrum the secondary electron BR dominates. The last fact is due to the "cut-off" frequency for the secondary electron BR being by 4 times larger than for non-coherent PBR.

Coherent proton PBR on the aluminium atom was calculated in [117] using Slater's wave functions for bound electrons and the corresponding screening constants. Two terms S_1 and S_2 contributed to the PBR amplitude. The first term (S_1) describes scattering of the IP proper field on the charge of the target's electronic core. The second term (S_2), vanishing in the high-frequency limit, includes the sum over intermediate states of the atomic energy spectrum. Note that in [117] S_2 was expressed in the closed analytical form. The calculation turned out to be in good agreement with experimental data [134] in the entire parameter range. The coherent PBR was found to contribute mainly to radiation of protons with energy 1 MeV. For protons with energy 4 MeV, the secondary electron BR is the dominating radiative mechanism.

To conclude this section, we concern ourselves with BR induced by heavy ion collisions with targets. Radiation from IP with energies from 7 to 18 MeV per nucleon was studied in experiments [137]. Multicharged ions N^{7+}, Ne^{10+}, Ar^{17+} were taken as incident particles and radiated in passing through various gaseous targets. The thickness of the gas layer was 6 mm and the gas pressure was normal. The experimental results obtained were interpreted in paper [138] for radiation of ions N^{7+} with an energy of 250 MeV in nitrogen and ions Ar^{17+} with an energy of 288 MeV in neon. The photon output was registered by the angle 90 degrees to the ionic beam axis in the spectral range from 4 keV to 20 keV. The high-frequency part of the neon ion emission spectrum (from 5 keV to 20 keV) was observed to have a shoulder-like form. In the argon ion emission spectrum inside the 4 – 12 keV range a sharp maximum was registered at the photon energy near 7 keV. To explain the observed spectra, in paper [138] contributions from three radiative mechanisms were taken into account: the radiative capture of the target's bound electrons in the IP continuum states, the radiative ionization (non-coherent PBR), and the secondary electron BR. Static BR, called nuclear in this paper, is negligibly small in this case. The described spectral structure was established as being due to contribution of the radiative capture of the target's electrons into the continuum of the IP energy states, which has the resonance-like spectral line form and is largest for the multicharged argon ion. The high-frequency "tail" in both cases is due to the secondary electron BR. The calculation also revealed the radiative ionization contribution to be significant for the nitrogen ion emission in the low-frequency part of the measured spectrum and to be negligible for the argon ion emission. This last fact is caused by the presence of the cut-off frequency in radiative ionization at $\omega_c = \mu v_i^2/2$, where μ is the reduced mass of electron and IP, v_i is the IP initial velocity (see Sect. 6.2).

10.4. Experiments on the laser-assisted electron scattering on atoms

When inelastic electron scattering on targets with a core occurs in the external electromagnetic field, a polarization mechanism mediating energy transfer from the field to the electron through a virtual excitation of the target can play an important role. This was predicted theoretically in paper [91]. As was said in the beginning of this Section, the inclusion of the polarization channel suggests an explanation to low thresholds of the laser break-down of the alkali metal vapors [99], which have a large polarizability of atoms.

There are other experiments on inelastic electron scattering on atoms in a laser field that have been carried out to measure energy spectra of scattered by a specific angle electrons. To interpret these experiments, calculation of the process cross section using static approximation proves to be insufficient in some cases.

One of the first papers of this kind [139] was devoted to studies of multiphoton processes in electron scattering on argon atoms assisted by an intensive CO_2 laser emission (with a peak power of 50 MW). The initial electron energy was taken 11 eV. It is essential that electrons were fixed by a large scattering angle (153 degrees). The experiment measured the number of scattered electrons with given energy. The laser field was found to significantly redistribute over energies the initially monoenergetic electron beam. The central peak corresponding to elastic scattering was observed to decrease by about 45%. At the same time, additional maxima appeared in the energy spectrum of scattered electrons, corresponding to absorption/emission of several laser photons up to $n = 3$. The obtained experimental data turned out to be in good consistence with predictions of a semi-classical theory elaborated in [140]. In particular, the so-called sum rule was shown to be fulfilled in the conditions of this experiment: the total scattering probability, summed over all photon outputs of the process, is constant. The target's polarization and statistical properties of the electromagnetic field effect on the sum rule for multiphoton induced BR was studied in paper [141] within the frames of the first Born approximation. It was shown that the allowance for the polarization channel essentially modifies the sum rule for sufficiently small electron scattering angles. The expression obtained in [141] for the total scattering cross section also entails that the role of polarization effects increases in transition from coherent to stochastic radiation. The contribution of these effects to the scattering transport cross section is maximal for the perpendicular orientation of the external field vector with respect to the initial electron velocity vector.

Paper [142], which continues experiments initiated in paper [139], reported on measurements of the electron spectra in scattering of electrons with energy 9.5 eV on the helium atoms by a small angle (9 degrees), assisted by a powerful ($P \approx 10^8$ W/cm^2) CO_2-laser emission. The additional electron peak intensities, corresponding to absorption/emission of the laser photons by electrons in the process of scattering, were found to significantly exceed values predicted by theory [140] which disregards the polarization channel. Experimental conditions were taken in [142] such that to satisfy as much as possible the applicability conditions of the so-call Kroll–Watson criterion [140] and to exclude excitations of the target's electronic core. The low polarizability of helium also was assumed to minimize polarization effects. Nevertheless, results [142] evidence for the Kroll–Watson formula being insufficient to explain the experimental data obtained.

Paper [143] studied theoretically the impact of the induced polarization of the target and statistical fluctuations of laser radiation (which in the case of a CO_2 laser is multimodal) on experimental data [142] for non-linear cross sections of free–free transitions. In particular, the possibility of helium atoms being in a metastable 2^1S-state for some reason with a polarizability of 803 a.u., i.e. by almost 600 times larger than that of the helium ground state 1^1S

(1.4 a.e.) was taken into account. To assess the role of polarization effects, formula (8.2) was used; this was obtained in [91] within the Born approximation frames and, as was noted in Sect. 8.1., does not take into account non-linear interaction of atomic electrons with electromagnetic field. As a result, a conclusion was made on polarization effects being negligible for the helium atom in the ground state. At the same time, for atoms in the metastable 2^1S state, accounting for the target's polarization strongly improved agreement with experiments [142]. However, a sharp decrease in the cross section with increasing photon output of the process observed in this case contradicts to experimental data. The stochasticity of multimode radiation of the CO_2 laser, taken into account in [143], somewhat smoothens this discrepancy. Nevertheless, the residual disagreement was used by the authors of paper [143] to draw a negative conclusion on the possibility to explain data [142] by making allowance for the polarization channel and statistical properties of laser radiation. It should be noted, however, that for the conclusion drawn to be completely correct, the evaluation of the role of polarization effects in multiphoton free – free transitions should be made using formula (8.3), taking into account non-linear interaction of the external field with atomic electrons.

10.5.   Collision-induced absorption in gases

There is a large amount of experimental material on the subject closely related to polarization effects in atomic transitions — viz., on absorption of electromagnetic radiation caused by collisions between molecules (collision-induced absorption — CIA) in gases. This problem is addressed in detail in fundamental monograph [144]. Here we briefly consider main properties of this phenomenon and describe some related experiments.

In contrast to ordinary (single-particle) absorption whose intensity is proportional to the first power of the particle number density (n), the collision-induced spectrum (I) is a non-linear function of the number density. The corresponding virial expansion for a low-pressure gas has the form

$$I = \tilde{A}\,n + \tilde{B}\,n^2 + \tilde{C}\,n^3 + \dots . \tag{10.1}$$

Here the first term describes the absorption of radiation by a non-perturbed atom or molecule and the remaining series — the collision-induced absorption. If the process is prevented for an individual molecule ($\tilde{A} = 0$), the contribution from CIA remains and can be significant for sufficiently high gas pressures.

Thus, collisions between gas particles lead to the appearance of absorption lines in the spectral ranges where there would be no absorption for isolated molecules. Non-linear terms in Eqn. (10.1) describe the corresponding binary, triple, etc., radiative collisions. Note that first experiments on this subject, carried out as early as in 1885 [145], discovered new absorption bands in oxygen under a pressure of tens and hundreds atmospheres, which were absent at the atmospheric pressure. The corresponding absorption coefficients proved to be proportional to the gas density square (violating the Bar's law), corresponding exactly to binary collisions.

According to modern concepts [144], the CIA spectra result from an electric dipole moment of the system, also called a super-molecule, appearing in the collision. From this viewpoint CIA is similar to PBR, in which the target's dipole moment is also induced by a collision with another particle. For PBR, in contrast to CIA, one of the particles must be charged. Note that PBR was considered for atom–atom collisions as well in paper [146].

A characteristic feature of the CIA spectra is that they have larger line widths than from single-particle absorption, usually of the order of $10^{12} - 10^{13}$ s^{-1}, which reflects the small life-time of the supermolecule. Another distinctive feature of the process under consideration is a small value of the induced dipole moment μ, which falls within the range $0.01 - 0.1$ Debye. Recall that for polar molecules the corresponding value is typically about several Debyes.

The most general expression for absorption of a quantum $\hbar\omega$ induced by a collision between particles A and B has the form

$$A + B + \hbar\omega \rightarrow A' + B' + \Delta\varepsilon. \qquad (10.2)$$

Here symbols A', B' denote the colliding molecules in excited states, $\Delta\varepsilon$ is the change in the translation energy of the particles. In the particular case $A = A'$, $B = B'$, all the photon energy is expired to increase the kinetic energy of the systems' translation motion, and the corresponding absorption spectrum is called translational [144]. This process is similar to the inverse polarization bremsstrahlung effect without target excitation. The translational spectrum is characteristic for CIA in single-atom inert gases, then the line center falls at . zero frequency. In other cases rotational and vibration – rotational molecular excitation occur as well, with transitions prohibited for isolated and excited molecules. The resulting spectrum is a vibration – rotation – translational with characteristic frequencies lying in far infrared and microwave wavelength range. The maximum frequencies of such a "rovibro – translational" spectrum correspond to energies of the vibration – rotation – translational transitions (an also their sums), with the band widths being determined by the supermolecule life time. Such spectra are universal in the sense that they appear in all molecular gases.

In some cases CIA can be accompanied by excitations of electronic degrees of freedom of the colliding particles. The corresponding spectrum lies in the visible and ultraviolet diapasons. CIA on electronic transitions $^3\Sigma_g^- \to a\,^1\Delta_g$, $b^1\Sigma_g^+$ in the oxygen molecule [147] may provide an example. Note that both these transitions in an isolated molecule are forbidden by selection rules. Nevertheless, experiment [147] reported on an absorption coefficient of about 24 cm^{-1} near the line maximum (7900 cm^{-1}), corresponding to the transition $^3\Sigma_g^- \to a\,^1\Delta_g$, measured at a temperature of 297 K and a gas particle number density of 132 amagats (1 amagat $\approx 2.7\,10^{19}$ cm^{-3}). Absorption bands corresponding to the excitation of the $b^1\Sigma_g^+$ state and simultaneous excitation of the colliding molecules to the $a^1\Delta_g$ and $b^1\Sigma_g^+$ states were also detected in this experiment.

CIA in a mixture of inert gases when there are no vibration – rotational degrees of freedom allows the most simple description. Note that CIA in a monoatomic gas is absent due to the absorbing supermolecule having a symmetry center. This makes the dipole moment impossible to emerge. The dipole moment in collision of two different atoms appears by two mechanisms: the exchange and dispersion ones [148]. The first mechanism typically operates at small distances R between the colliding particles and the corresponding dipole moment falls off exponentially with R, $\mu_O \approx \mu_0 \exp\left(-R/R_0\right)$, here R_0 is the distance of the maximum approach. The dispersion mecha-

nism is essential at large R. Then the induced dipole moment admits
the expansion $\mu_D \approx -D_7/R^7 - D_9/R^9 - \ldots$ [144].

If at least one of the colliding particles is a (non-monoatomic) molecule, other mechanisms for the dipole momentum appearance can operate, which are related to the far-acting electric field of the molecule. For symmetric diatomic molecules like H_2 or N_2, the lowest term of the molecular field expansion is quadrupole, and the dipole moment induced by this field in the particle–partner at large distances can be represented in the form

$$\mu_Q \approx \sqrt{3}\, q_2\, \alpha \Big/ R^4, \quad R \gg R_0 . \qquad (10.3)$$

Here q_2 is the molecule's quadrupole moment, α is the particle–partner's polarizability. Equality (10.3) revealed the mentioned above analogy between CIA and polarization effects in atomic transitions: in both cases the interaction with electromagnetic field is mediated by polarization of the electronic core of one of the colliding particles by the field of the particle–partner. The induced dipole moment can be induced by gradients of the electric molecular field and also by a collisional breaking of the initial molecular symmetry.

The spectral absorption coefficient in an inert gas mixture due to binary collisions measured at the room temperature [149, 150] is presented in Fig. 35. The absorption coefficient $\kappa_a(\nu)$ is normalized on the product of the component number densities, which allowed us to show in one plot the data obtained for different pressures in mixtures. A characteristic feature of the demonstrated relationships is the presence of broad maxima in the far infrared diapason that shift towards high frequencies as molecular velocity increases. Using data from Fig. 35

$$\kappa_a(\nu) \propto n_1\, n_2 \nu\, g(\nu) \left\{ \nu \left[1 - \exp\left(-\frac{hc\nu}{kT} \right) \right] \right\}, \qquad (10.4)$$

where k is the Boltzmann constant, c is the speed of light, T is the temperature, one can recover spectral functions $g(\nu)$ that describe CIA in different gas mixtures. Getting rid off the frequency-dependent factor in the curved brackets of Eqn. (10.4) results in the maximum of the spectral function $g(\nu)$ falling at the zero frequency,

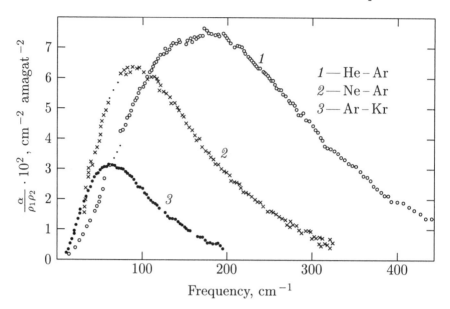

Fig. 35. The results of measurement of the binary absorption coefficient by inertial gas mixtures He – Ar, Ne – Ar [149], and Ar – Kr [150] at the room temperature, normalized to the product of concentrations.

which is typical for the translational CIA spectra. The spectral function half-width can be evaluated using the formula

$$\Delta\nu = \frac{2.5}{\pi\,c\,\sigma}\,\bar{v}_{12}\,, \qquad (10.5)$$

where $\bar{v}_{12}$ is the mean relative velocity of different atoms, the parameter σ is the square root of the interatom interaction potential $V_{\mathrm{int}}(\sigma) = 0$. This quantity for mixtures He – Ne, Ne – Ar, Ar – Kr takes the value 0.295, 0.305, 0.345 nm, respectively. The spectral function half-widths for the given mixtures are obtained from Eqn. (10.5) to be 129, 60, 37 cm^{-1}, respectively, which is in reasonable agreement with experimental data from Fig. 35. Note that the form of the spectral function considered can be approximated by the Lorentzian curve at the center (i.e. near the zero frequency) and falls off exponentially at the edges.

An interesting feature of the CIA spectra, which has a direct analogy with the density effect for the relativistic particle PBR in an amorphous condensed medium (see Sect. 6.3), is the presence of the so-called "inter-collisional dips" in the spectral function and the absorption coefficient at sufficiently low frequencies. For example, the spectral function of CIA in a 1 : 1 mixture of helium and argon at the total pressure of 160 atm reveals a dip at frequencies below 10 cm^{-1}. This effect results from a destructive interference of contributions to absorption due to successive collisions of gas molecules, when the time interval between the collisions becomes of the order of the inverse frequency of the electromagnetic field. The destructive interference appears because the induced dipole moments in two successive collisions between molecules turn out to be more or less antiparallel, like a fast electron in a condensed medium induces antiparallel polarization in the medium's atoms on both side of its trajectory.

CIA plays an important role in astrophysical applications, especially when other radiative mechanisms do not contribute to the considered spectral range, for example in cool areas of planet atmospheres. For example, the first direct evidence of the presence of H_2 molecules in atmospheres of the external planets was obtained in [151] by reproducing in the laboratory conditions a diffusive structure at the wavelength of 827 nm observed in spectra of Uranus and Neptune. Experiment [151] studied radiation absorption in hydrogen at the 80-m wavelength and a temperature of 73 K. This structure proved to be the line $S_3(0)$ in the collision-induced vibration–rotational band $0 \rightarrow 3$ of the H_2 molecule. In this paper, another spectral feature at the wavelength of 816.6 nm was identified to be a collision-induced double transition in H_2 molecules. The experiment used a pure hydrogen and the observed double transition was relatively strong. At the same time, it was observed to be much weaker than the line $S_3(0)$ in spectra of the external planets. This fact enabled making the conclusion that an appreciable amount of helium is present in the atmospheres of Uranus and Neptune as helium notably strengthens the $S_3(0)$ line and cannot affect the double transition in colliding hydrogen atoms. Presently, CIA is recognized to play an important role in the temperature balance and atmospheric structure of giant planets. The CIA-spectra are also interesting for studies of some stellar atmospheres, such as in late stars, low mass stars, brown dwarfs, some cold white dwarfs, etc.

In conclusion, we point to another relative CIA-phenomenon that can also be interpreted to be a polarization effect, viz., an atom excitation in collision with a photon and an electron [152]. This is a direct analog of CIA with the substitution of the colliding neutral particles by electron. It can also be represented using Feynmann diagrams, shown in the Introduction, and its cross section can be expressed through the non-diagonal matrix element of the operator of the radiation scattering on atom. This process was studied in detail experimentally [153] using the technique of triple collision of electron – photon – atomic beams. In particular, the excitation of the $2(^3S)$ state of the helium atom was observed in collision with electron assisted by a low-power CO_2 laser field in the form of two satellite peaks in the spectrum of energy losses of inelastically scattered electrons at $E = 19.817 \pm 0.117$ eV.

A multiphoton modification of atomic excitations in simultaneous collision with electron and photon was studied theoretically in [154].

10.6. Polarization effects near the $4f$-structure in BR on metallic targets

Experiments [155, 156] examined emission spectra of M-series of the metallic lanthanum and cerium (the wavelength $14-15$ A) excited by an electronic beam with the near-threshold energies (the current up to 100 mA). As a result, a spectral structure (called $4f$-structure) was discovered, whose amplitude – frequency characteristics were in close relation with the energy of activating electrons. An important signature of this structure was a drastic intensity increase in the electronic beam energy approaching the excitation potentials of the $3d_{5/2}$, $3d_{3/2}$ subshells of lanthanum and cerium. An analysis of a large amount of emission spectra (40 for lanthanum and 70 for cerium), registered in changes within the electronic beam energy in $1-2$ eV steps with the energy dispersion of less than 0.2 eV. This revealed that the discovered structure is a peculiarity in the continuum emission spectrum of exciting electrons with their transition to the

$4f$-state located above the Fermi sea, and not the usual character-istic emission of M-lines of the corresponding atoms. This structure was also observed for the beam energies away from the excitation potential of the $3d$-subshell in the form of a peak shifted from the short-wavelength boundary of BR toward low energies (by 5.5 eV for lanthanum). Such a spectral localization of the emission maximum is consistent with data on the energy of the $4f$-state lying by 5.5 eV above the Fermi level in the metallic lanthanum. The intensity of this peak decreased with electron energies up to the detuning off the M-series excitation potential by $10-15$ eV, while the location with respect to the short-wavelength boundary remained unchanged.

For the beam energy approaching the energy of the charac-teristic M_α-line ($E_{MV}(La) = 834$ eV), the radiation intensity at the $4f$-structure center ($\hbar\omega_{max} \simeq 831$ eV) increased by more than two orders of magnitude, approaching a maximum (for lanthanum) at $E_{el.beam} = 836.5$ eV. Further intensity decrease in the structure center with excitation energy growth proceeded more slow than its increase in the low-frequency wing (for $\hbar\omega_{max} < E_{MV}$), until the $4f$-structure central radiation intensity increased again near the energy of the M_β-line photons. Thus a spectral asymmetry in the emission $4f$-structure excitation was observed: the radiation intensity in the low-frequency wing of the line was lower than in the high-frequency wing.

Figure 36, taken from paper [156], shows the dependence of the radiation intensity in the $4f$-structure on its central frequency for cerium. The minimum of the intensity was observed at a photon energy of 868.1 eV. The radiation intensity in this minimum turned out to be 175 times as small as in the first maximum at 882.2 eV and 105 times as low as the second maximum intensity at 900.2 eV. The peak intensity ratio $5:3$ is approximately equal to the statisti-cal weight ratio for M_5 and M_4-levels, and their central frequencies are somewhat shifted toward low energies with respect to frequencies of the M_α and M_β lines. This shift has the opposite sign as the corresponding shift of the maximum radiation near the $4d$-subshell ionization potential with respect to photoabsorption maximum in lan-thanum [121] and xenon [120]. Figure 36 also clearly demonstrates the asymmetry in the spectral form of the $4f$-structure mentioned above.

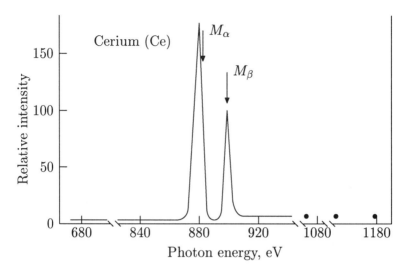

Fig. 36. The intensity of the bremsstrahlung with electron capture into the $4f$-state in scattering on metallic cerium as a function of the electron beam energy [156].

In the experimental works cited above, a qualitative interpretation of the obtained results was suggested based on the assumption that the increase in the inelastic scattering cross section of the emitting electron into the $4f$-state is related to the formation of a short-living excited state of negative ion $3d^{-1}4f^2$. It was stressed then that for atoms of the targets considered, a strong overlapping of the $4f$-wave functions with the $3d$-wave functions takes place, which provides a strong spatial coupling between the corresponding states.

Asymmetric resonances in BR observed in experiments [155, 156] were reproduced theoretically in paper [157] using the notion of the polarization BR mechanism. This paper in particular stressed that the important role of the polarization channel in considered cases was due to a large polarizability of the $3d$-subshells of lanthanum and cerium atoms. Qualitatively, the picture is as follows. The scattering electron, in addition to emission due to the proper dipole moment, excites by its Coulomb field an oscillating dipole moment on the atomic transition $(3d^9 4f) - (3d^{10})$, which ultimately leads to a resonance increase in the $4f$-structure intensity. Within this approach, the total emitting dipole moment is the sum of two terms — the direct and polarization — in correspondence with the expression [157]

$$D(E) \propto \langle 4f|z|E_g \rangle - \frac{\langle 4f4f|1/r_{21}|3dEg \rangle \langle 3d|z|4f \rangle}{E_{4f} + E_{4f} - E_{3d} - i\,\Gamma/2 - E} + \text{exchange terms}.$$
(10.6)

Here E is the incident electron energy, E_g is the g-component of its wave function, Γ is the width of the resonance state of the negative ion ($3d^{-1}4f^2$), z is the projection of the incident electron's radius vector. For the relative intensity, which is an analog to the spectral R-factor, the following expression was obtained:

$$R(E) = \frac{I(E)}{I_0(E)} = \left| 1 + \sum_{i=1}^{2} \frac{a_i(E)}{E - E_i + i\,\Gamma/2} \right|^2.$$
(10.7)

Here functions $a_i(E) = g_i\,b(E)$, are introduced, where g_i are statistical weights of the states and $b(E)$ is a coupling constant determined through the corresponding radial integrals. The coupling constant was calculated from the first principles to be $b(E = 62\ \text{Ry}) = 1.1\ \text{Ry}$, which was in a reasonable agreement with experimental values $b = 10 - 15$ eV [155].

With the use of Eqn. (10.6), the $4f$-peak intensity was computed based on the dipole formula $I(E) \propto |D(E)|^2$ with allowance for the $3d$-subshell fine splitting. For crude comparison with experiments on lanthanum, the following parameter values were chosen in this paper: the resonance width $\Gamma = 0.2$ Ry, the fine splitting value $E_2 - E_1 = 1.2$ Ry. The model calculations successfully reproduced the main features of the phenomenon found by experiment, such as the asymmetry of the emission resonances with interference dip in the low-frequency line wing. Besides, the calculated energy detuning of the interference minimum off the first maximum position proved to be equal to 10 eV. This is consistent with experimental values ranging from $10 - 15$ eV [135]. In addition, the calculated radiation intensity excess in the $4f$-peaks over the background intensity turned out to be somewhat lower than the experimentally registered value.

The considered effect of resonant intensity increase in radiation of the $4f$-structure near the $3d$-subshell ionization potential does not occur for frequencies corresponding to excitation of the $4d$-electrons, as was shown, in particular, by calculations [158]. As was noted in paper [106], this fact is apparently due to a low formation probability of short-living dielectronic states ($4d^9 4f^2$) in lanthanum.

11. Conclusion

The above consideration suggests that the plasma models of atom provides a fairly reasonable approximation to consider a wide range of radiation – collisional effects with participation of heavy atoms and ions. The plasma approach is advantageous in being universal and providing possibility to establish various similarity laws ("scalings") for the effects under study. Naturally, the averaged statistical description does not take into account some peculiarities of individual atomic spectra. As a rule, these features are manifest in specially designed experiments involving directed beams of charged particles when one can ignore effects of the ambient medium. At the same time the averaged radiation characteristics, satisfactorily described by the statistic plasma models, are important for radiation of particles in the medium. In particular, the application of the statistic approach allows one to calculate BR on atoms in a wide range of photon energy [159], as well as to compute bremsstrahlung emission and radiative recombination rates from different chemical element admixtures in a high-temperature plasma.

There is a great variety of polarization radiation effects considered above including radiative transitions of different types. The intensity of polarization radiation of electrons on complex atoms and ions varies within a broad range determined by the generalized target's polarizability and effective ion charge acting on electron and changing from the ion's charge to the nuclear charge. The radiation can be coherent or non-coherent with respect to the contribution of atomic electrons, depending on the value of the momentum transmitted to the target.

In practical applications of special interest are integral characteristics of PR, such as the total radiative losses in plasma with heavy ions. The contribution of the polarization channel can be naturally characterized by the integral R-factor (R_{tot}), which is the ratio of the integral over frequency polarization losses to integral losses in the

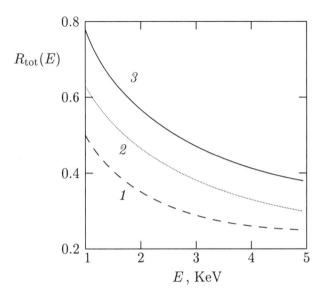

Fig. 37. The integral R-factor dependence on the incident electron energy E for different nuclear charges of the Thomas – Fermi atom Z: $1 - Z = 30$, $2 - Z = 60$, $3 - Z = 90$.

static channel. The dependence of the factor R_{tot} on the incident electron energy for different nuclear charges is shown in Fig. 37. It is seen that the polarization channel contribution is comparable to that from the static channel, especially for mild electron energies. With increasing energy of electrons, their penetration into the core increases and the polarization channel contribution diminishes [161].

Of a great interest are polarization effect studies in line spectra of multicharged ions that appears as a set of bound – bound transition lines corresponding to states with a complex electronic configuration of the core. With increasing the ionic charge, i.e. the radiative transition frequency, the role of the dynamic polarizability of the core should increase, in contrast to the static polarizability, again characteristic to transitions in the alkali elements.

Direct measurements of the polarization recombination are certainly interesting, too. Apparently, this can be most simply carried out in experiments with multicharged ions using storage rings. Here, however, the polarization recombination observations are related to a thus far unclear reason for a significant increase in the total recombination probability. This was also found for ions without any electronic core [66].

The polarization radiation from heavy particles, in particular, for multicharged ions in their motion inside dense gaseous or solid targets, is studied to a much less degree than the polarization radiation from electrons. The main problem in studying these phenomena lies in distinguishing between the proper polarization radiation and the radiation from secondary electrons resulted from the target's matter ionization.

The polarization radiation, including the polarization recombination, is an essential effect for the inelastic electron scattering on metallic clusters near the giant resonance frequency [162]. Radiation properties of the clusters have been studied by the present time quite well (see monograph [163]). They can be used to calculate polarization characteristics of these objects and the corresponding PR intensities (see review [164]). Studies of such phenomena on nanoparticles of other types, such as semi-conductor and dielectric nanocrystals, seem to be very interesting.

Interference effects in the inelastic electron scattering on ions with a core caused by the interaction of the usual and polarization channels, studied in detail theoretically [93, 102], need to be verified experimentally. The prospects of their practical applications should also be investigated.

Significantly interesting are studies of radiation from hot gases appeared in shock waves or cavitation waves in sonoluminescence. BR has been considered as a possible mechanism for emission in these phenomena [165]. Here the conditions may be made such that PR prevails over the static channel due to the elastic electron scattering [144, 166].

The question on the role of polarization radiation in a cold medium is very interesting and remains to be solved. Here we deal with the emission from a cold medium in which there are virtually no internal degrees of freedom capable of producing radiation. In these conditions only the polarization channel remains that can provide radiative cooling at any finite temperature. The possibility of performing the corresponding experiments needs to be additionally justified.

Thus we can state that the notion of atom as a plasma clot and the related treatment of the polarization radiation, in addition to its important methodological meaning, has also proven very fruitful in practical applications.

The authors deeply acknowledge V. M. Buimistrov, V. I. Kogan, A. B. Kukushkin, B. M. Smirnov, and A. N. Starostin for valuable discussions.

The work was done with the financial support from the Russian Fund for Basic Research (the project No 01 – 02 – 16305), from the MNTC grant No 2155, and from the Ministry of Science under the project "Fundamental Spectroscopy".

REFERENCES

1. P. Gombás, *Die statistische Theorie des Atoms und ihre Anwendungen*, Springer-Verlag, Wien (1949). Russian translation: *Statisticheskaya Teoriya Atoma i ee Primeneniya* (Statistic Atomic Theory and Its Applications), IL, Moscow (1951)

2. L. D. Landau and E. M. Lifshitz, *Kvantovaya Mekhanika*: *Nerelyativistskaya Teoriya* (Quantum Mechanics: Non-Relativistic Theory) [in Russian], Nauka, Moscow (1974). English translation: Pergamon Press, Oxford (1977)

3. W. Brandt and S. Lundqvist, *Phys. Rev.*, **139**, A612 (1965)

4. *Theory of the Inhomogeneous Electron Gas* (edited by S. Lundqvist, N. H. March), Plenum Press, New York (1983). Russian translation: Mir, Moscow (1987)

5. A. V. Vinogradov and V. P. Shevel'ko, *Tr. Fiz. Inst. Akad. Nauk SSSR*, **119**, 158 (1980)

6. A. V. Vinogradov and O. I. Tolstikhin, *Zh. Eksp. Teor. Fiz.*, **96**, 61 (1989). English translation: *Sov. Phys. JETP*, **69**, 32 (1989)

7. A. V. Vinogradov and O. I. Tolstikhin, *Zh. Eksp. Teor. Fiz.*, **96**, 1204 (1989). English translation: *Sov. Phys. JETP*, **69**, 683 (1989)

8. A. D. Ulantsev and V. P. Shevelko, *Opt. Spektrosk.*, **65**, 1003 (1988). English translation: Opt. Spectrosc., **65**, 590 (1988)

9. *Polyarizatsionnoe Tormoznoe Izluchenie Chastits i Atomov* (Polarization Bremsstrahlung of Particles and Atoms) [in Russian]

(edited by V. N. Tsytovich and I. M. Oiringel'), Nauka, Moscow (1987). English translation: *Polarization Bremsstrahlung* (edited by V. N. Tsytovich and I. M. Oiringel'), Plenum Press, New York (1992)

10. A. V. Korol and A. V. Solov'yov, *J. Phys. B: At. Mol. Opt. Phys.*, **30**, 1105 (1997)

11. V. A. Astapenko, L. A. Bureeva, and V. S. Lisitsa, *Usp. Fiz. Nauk*, **172**, 155 (2002). English translation: *Phys. Usp.*, **45**, 149 (2002)

12. V. D. Shafranov, in: *Voprosy Teorii Plazmy* (Topics on Plasma Theory) [in Russian], (edited by M. A. Leontovich), 3rd ed., Issue 3, Gosatomizdat, Moscow (1963), p. 3

13. V. L. Ginzburg and V. N. Tsytovich, *Perekhodnoe Izluchenie i Perekhodnoe Rasseyanie* (Transition Radiation and Transition Scattering) [in Russian], Nauka, Moscow, (1984). English translation: A. Hilger, Bristol (1990)

14. G. V. Gadiyak, D. A. Kirzhnits, and Yu. E. Lozovik, *Zh. Eksp. Teor. Fiz.*, **69**, 122 (1975)

15. M. Ya. Amusia, *Atomnyi Fotoeffekt* (Atomic Photoeffect) [in Russian], Nauka, Moscow (1987). English translation: Plenum Press, New York (1990)

16. Yu. L. Klimontovich, *Statisticheskaya Fizika* (Statistical Physics) [in Russian], Nauka, Moscow (1982). English translation: Harwood Acad. Publ., Chur (1986)

17. L. Bureyeva and V. Lisitsa, *J. Phys. B: At. Mol. Opt. Phys.*, **31**, 1477 (1998)

18. N. F. Mott and H. S. W. Massey, *The Theory of Atomic Collisions*, 3rd ed., Clarendon Press, Oxford (1965). Russian translation: Mir, Moscow (1969)

19. S. P. Kapitza, *Zh. Eksp. Teor. Fiz.*, **39**, 1367 (1960). English translation: *Sov. Phys. JETP*, **12**, 943 (1961)

20. M. L. Ter-Mikaelyan, *Dokl. Akad. Nauk SSSR*, **134**, 318 (1960)

21. L. D. Landau and E. M. Lifshitz, *Elektrodinamika Sploshnykh Sred* (Electrodynamics of Continuous Media) [in Russian], Nauka, Moscow (1982). English translation: Pergamon Press, Oxford (1984)

22. E. Fermi, *Z. Phys.*, **29**, 315 (1924)

23. P. M. Bergstrom, Jr. and R. H. Pratt, *Radiat. Phys. Chem.*, **50**, 3 (1997)

24. A. Dubois and A. Maquet, *Phys. Rev. A*, **40**, 4288 (1989)

25. V. A. Astapenko, L. A. Bureeva, and V. S. Lisitsa, *Zh. Eksp. Teor. Fiz.*, **117**, 496 (2000). English translation: *JETP*, **90**, 434 (2000)

26. M. Ya. Amus'ya, N. A. Cherepkov, and S. G. Shapiro, *Zh. Eksp. Teor. Fiz.*, **63**, 889 (1972). English translation: *Sov. Phys. JETP*, **36**, 554 (1973)

27. M. J. Stott and E. Zaremba, *Phys. Rev. A*, **21**, 12 (1980)

28. A. Zangwill and P. Soven, *Phys. Rev. A*, **21**, 1561 (1980)

29. W. Brandt, L. Eder, and S. Lundqvist, *J. Quant. Spectrosc. Radiat. Transf.*, **7**, 185 (1967)

30. A. V. Korol', A. G. Lyalin, O. I. Obolenskii, and A. V. Solov'jev, *Zh. Eksp. Teor. Fiz.*, **114**, 458 (1998). English translation: *JETP*, **87**, 251 (1998)

31. B. Edlén, *Phys. Scripta*, **17**, 565 (1978)

32. L. P. Rapoport, B. A. Zon, and N. L. Manakov, *Teoriya Mnogofotonnykh Protsessov v Atomakh* (Theory of Many-Photon Processes in Atoms) [in Russian], Atomizdat, Moscow (1978)

33. V. M. Buimistrov and L. I. Trakhtenberg, *Zh. Eksp. Teor. Fiz.*, **69**, 108 (1975). English translation: *Sov. Phys. JETP*, **42**, 55 (1975)

34. P. A. Golovinskii and B. A. Zon, *Opt. Spektrosk.*, **45**, 854 (1978). English translation: *Opt. Spectrosc.*, **45**, 733 (1978)

35. R. R. Teachout and R. T. Pack, *Atom. Data*, **3**, 195 (1971)

36. N. B. Delone and V. P. Krainov, *Zh. Eksp. Teor. Fiz.*, **83**, 2021 (1982). English translation: *Sov. Phys. JETP*, **56**, 1170 (1982)

37. I. L. Beigman, *Zh. Eksp. Teor. Fiz.*, **100**, 125 (1991). English translation: *Sov. Phys. JETP*, **73**, 68 (1991)

38. J. D. Bjorken and S. D. Drell, *Relativistic Quantum Mechanics*, McGraw-Hill, New York (1964). Russian translation: *Relyativistskaya Kvantovaya Teoriya* (Relativistic Quantum Theory), Vol. 1, *Relyativistskaya Kvantovaya Mekhanika* (Relativistic Quantum Mechanics), Mir, Moscow (1978)

39. J. A. Wheeler and W. E. Lamb, Jr., *Phys. Rev.*, **55**, 858 (1939)

40. V. L. Ginzburg, *Teoreticheskaya Fizika i Astrofizika* (Theoretical Physics and Astrophysics) [in Russian], Nauka, Moscow (1987). English translation: *Applications of Electrodynamics in Theoretical Physics and Astrophysics*, Gordon and Breach, New York (1989)

41. D. Pines, *Elementary Excitations in Solids*, W. A. Benjamin, New York (1963). Russian translation: Mir, Moscow (1965)

42. P. M. Platzman and P. A. Wolff, *Waves and Interactions in Solid State Plasmas*, Academic Press, New York (1973). Russian translation: Mir, Moscow (1975)

43. J. D. Jackson, *Classical Electrodynamics*, 2nd ed., Wiley, New York (1975). Russian translation: Mir, Moscow (1979)

44. L. D. Landau and E. M. Lifshitz, *Teoriya Polya* (The Classical Theory of Fields) [in Russian], Nauka, Moscow (1973). English translation: Pergamon Press, Oxford (1975)

45. V. I. Gervids and V. I. Kogan, *Pis'ma Zh. Eksp. Teor. Fiz.*, **22**, 308 (1975). English translation: *JETP Lett.*, **22**, 150 (1975)

46. V. I. Kogan and A. B. Kukushkin, *Zh. Eksp. Teor. Fiz.*, **87**, 1164 (1984). English translation: *Sov. Phys. JETP*, **60**, 665 (1984)

47. V. I. Kogan, A. B. Kukushkin, and V. S. Lisitsa, *Phys. Rep.*, **213**, 1 (1992)

48. H. A. Kramers, *Philos. Mag.*, **46**, 836 (1923)

49. J. M. Rost, *J. Phys. B: At. Mol. Opt. Phys.*, **28**, L601 (1995)

50. P. A. Golovinskii and I. Yu. Kiyan, *Usp. Fiz. Nauk*, **160**, No. 6, 97 (1990). English translation: *Sov. Phys. Usp.*, **33**, 453 (1990)

51. C. Blondel, M. Crance, and C. Delsart, et al., *J. Phys. II*, **2**, 839 (1992)

52. P. A. Golovinskii and B. A. Zon, *Zh. Tekh. Fiz.*, **50**, 1847 (1980). English translation: *Sov. Phys. Tech. Phys.*, **25**, 1076 (1980)

53. A. A. Radzig and B. M. Smirnov, *Parametry Atomov i Atomnykh Ionov* (Parameters of Atoms and Atomic Ions) [in Russian], Energoatomizdat, Moscow (1986)

54. V. D. Kirillov, B. A. Trubnikov, and S. A. Trushin, *Fiz. Plazmy*, **1**, 218 (1975). English translation: *Sov. J. Plasma Phys.*, **1**, 117 (1975)

55. V. P. Zhdanov and M. I. Chibisov, *Zh. Tekh. Fiz.*, **47**, 1804 (1977)

56. V. P. Zhdanov, *Fiz. Plazmy*, **4**, 128 (1978). English translation: *Sov. J. Plasma Phys.*, **4**, 71 (1978)

57. C. M. Lee, L. Kissel, R. H. Pratt, and H. K. Tseng, *Phys. Rev. A*, **13**, 1714 (1976)

58. R. Hippler, K. Saeed, I. McGregor, and H. Kleinpoppen, *Phys. Rev. Lett.*, **46**, 1622 (1981)

59. B. A. Zon, *Zh. Eksp. Teor. Fiz.*, **77**, 44 (1979). English translation: *Sov. Phys. JETP*, **50**, 23 (1979)

60. V. A. Astapenko, *AIP Conf. Proc.*, **559**, 176 (2001)

61. V. A. Astapenko, L. A. Bureeva, and V. S. Lisitsa, *Zh. Eksp. Teor. Fiz.*, **117**, 906 (2000). English translation: *JETP*, **90**, 788 (2000)

62. V. V. Ivanov, A. B. Kukushkin, and V. I. Kogan, *Fiz. Plazmy*, **15**, 1531 (1989)

63. V. A. Kas'yanov and A. N. Starostin, *Zh. Eksp. Teor. Fiz.*, **48**, 295 (1965). English translation: *Sov. Phys. JETP*, **21**, 192 (1965)

64. V. A. Astapenko, *Laser Phys.*, **11**, 1042 (2001)

65. V. A. Astapenko, L. A. Bureeva, and V. S. Lisitsa, *Zh. Eksp. Teor. Fiz.*, **121**, 19 (2002). English translation: *JETP*, **94**, 12 (2002)

66. A. Müller, *Nucl. Instrum. Meth. B*, **99**, 58 (1995)

67. K. Dick and H. J. Pepin, *Appl. Phys.*, **44**, 3284 (1973)

68. D. E. Post, R. V. Jensen, C. B. Tarter, et al., *Atom. Data Nucl. Data Tabl.*, **20**, 397 (1977)

69. S. N. Nahar, *Phys. Rev. A*, **55**, 1980 (1997)

70. D. T. Woods, J. M. Shull, and C. L. Sarazin, *Astrophys. J.*, **249**, 399 (1981)

71. M. Arnaud, J. Raymond, *Astrophys. J.*, **398**, 394 (1992)

72. V. A. Astapenko, *Fiz. Plazmy*, **27**, 503 (2001). English translation: *Plasma Phys. Rep.*, **27**, 474 (2001)

73. A. V. Akopyan and V. N. Tsytovich, *Fiz. Plazmy*, **1**, 673 (1975). English translation: *Sov. J. Plasma Phys.*, **1**, 371 (1975)

74. A. Messiah, *Mécanique Quantique*, Dunod, Paris (1959). English translation: *Quantum Mechanics*, North-Holland Publ.

Co., Amsterdam (1961). Russian translation: Vol. 1, Nauka, Moscow (1978)

75. A. V. Akopyan and V. N. Tsytovich, *Zh. Eksp. Teor. Fiz.*, **71**, 166 (1976). English translation: *Sov. Phys. JETP*, **44**, 87 (1976).

76. V. A. Astapenko, V. M. Buimistrov, Yu. A. Krotov, and V. N. Tsytovich, *Fiz. Plazmy*, **15**, 202 (1989). English translation: *Sov. J. Plasma Phys.*, **15**, 116 (1989)

77. V. M. Buimistrov, L. I. Trakhtenberg, *Zh. Eksp. Teor. Fiz.*, **73**, 850 (1977). English translation: *Sov. Phys. JETP*, **46**, 447 (1977)

78. V. A. Astapenko, in: *Fizicheskie Yavleniya v Priborakh Elektronnoi i Lazernoi Tekhniki* (Physical Phenomena in Instruments of Electronic and Laser Techniques) [in Russian], MFTI, Moscow (1985), p. 55

79. N. N. Nasonov and A. G. Safronov, *Zh. Tekh. Fiz.*, **62**, 1 (1992)

80. K. Yu. Platonov and I. N. Toptygin, *Zh. Eksp. Teor. Fiz.*, **98**, 89 (1990). English translation: *Sov. Phys. JETP*, **71**, 48 (1990)

81. M. L. Ter-Mikaelyan, *Vliyanie Sredy na Elektromagnitnye Protsessy pri Vysokikh Energiyakh* (Medium Effect on Electromagnetic Processes at High Energies) [in Russian], Izd. Akad. Nauk Arm. SSR, Erevan (1969). English translation: *High-Energy Electromagnetic Processes in Condensed Media*, Wiley-Intersci., New York (1972)

82. N. N. Nasonov, *Nucl. Instrum. Meth. B*, **145**, 19 (1998)

83. O. B. Firsov and M. I. Chibisov, *Zh. Eksp. Teor. Fiz.*, **39**, 1770 (1960). English translation: *Sov. Phys. JETP*, **12**, 1232 (1961)

84. T. Ohmura and H. Ohmura, *Phys. Rev.*, **121**, 513 (1961)

85. M. Ya. Amus'ya, A. S. Baltenkov, and A. A. Paiziev, *Pis'ma Zh. Eksp. Teor. Fiz.*, **24**, 366 (1976). English translation: *JETP Lett.*, **24**, 332 (1976)

86. H. S. W. Massey and E. H. S. Burhop, *Electronic and Ionic Impact Phenomena*, Clarendon Press, Oxford (1952). Russian translation: IL, Moscow (1958)

87. V. B. Berestetskii, E. M. Lifshitz, and L. P. Pitaevskii, *Kvantovaya Elektrodinamika* (Quantum Electrodynamics) [in Russian], Nauka, Moscow (1989). English translation: Pergamon Press, Oxford (1982)

88. N. B. Delone and V. P. Krainov, *Nelineinaya Ionizatsiya Atomov Lazernym Izlucheniem* (Nonlinear Ionization of Atoms by Laser Radiation) [in Russian], Fizmatlit, Moscow (2001)

89. F. V. Bunkin and M. V. Fedorov, *Zh. Eksp. Teor. Fiz.*, **49**, 1215 (1965). English translation: *Sov. Phys. JETP*, **22**, 844 (1966)

90. I. Ya. Berson, *Zh. Eksp. Teor. Fiz.*, **80**, 1727 (1981). English translation: *Sov. Phys. JETP*, **53**, 854 (1981)

91. B. A. Zon, *Zh. Eksp. Teor. Fiz.*, **73**, 128 (1977). English translation: *Sov. Phys. JETP*, **46**, 67 (1977)

92. P. A. Golovinskii, *Zh. Eksp. Teor. Fiz.*, **94**, No. 7, 87 (1988). English translation: *Sov. Phys. JETP*, **67**, 1346 (1988)

93. V. A. Astapenko and A. B. Kukushkin, *Zh. Eksp. Teor. Fiz.*, **111**, 419 (1997). English translation: *JETP*, **84**, 229 (1997)

94. V. S. Lisitsa and Yu. A. Savel'ev, *Zh. Eksp. Teor. Fiz.*, **92**, 484 (1987). English translation: *Sov. Phys. JETP*, **65**, 268 (1987)

95. P. A. Golovinskii and M. A. Dolgopolov, *Teor. Mat. Fiz.*, **95**, 418 (1993). English translation: *Theor. Math. Phys.*, **95**, 686 (1993)

96. A. A. Makarov, *Zh. Eksp. Teor. Fiz.*, **85**, 1192 (1983). English translation: *Sov. Phys. JETP*, **58**, 745 (1983)

97. V. A. Kas'yanov and A. N. Starostin, *Zh. Eksp. Teor. Fiz.*, **76**, 944 (1979). English translation: *Sov. Phys. JETP*, **49**, 654 (1979)

98. V. A. Kas'yanov and A. N. Starostin, *Kvantovaya Elektron.*, **8**, 1059 (1981)

99. J. E. Rizzo and R. C. Klewe, *Brit. J. Appl. Phys.*, **17**, 1137 (1966)

100. L. I. Trakhtenberg, *Opt. Spektrosk.*, **44**, 863 (1978). English translation: *Opt. Spectrosc.*, **44**, 510 (1978)

101. G. Kracke, J. S. Briggs, A. Dubois, et al., *J. Phys. B: At. Mol. Opt. Phys.*, **27**, 3241 (1994)

102. V. A. Astapenko, *Zh. Eksp. Teor. Fiz.*, **115**, 1619 (1999). English translation: *JETP*, **88**, 889 (1999)

103. V. A. Astapenko, *Laser Phys.*, **9**, 1032 (1999)

104. V. F. Bunkin, A. E. Kazakov, and M. V. Fedorov, *Usp. Fiz. Nauk*, **107**, 559 (1972)

105. F. Biggs, L. B. Mendelsohn, and J. B. Mann, *Atom. Data Nucl. Data Tabl.*, **16**, 201 (1975)

106. M. Gryzinski, *Phys. Rev.*, **138**, A336 (1965)

107. W. Lotz, *Z. Phys.*, **206**, 205 (1967)

108. B. M. Smirnov, *Iony i Vozbuzhdennye Atomy v Plazme* (Ions and Excited Atoms in Plasma) [in Russian], Atomizdat, Moscow (1974)

109. Y.-K. Kim and M. E. Rudd, *Phys. Rev. A*, **50**, 3954 (1994)

110. R. S. Freund, R. C. Wetzel, R. J. Shul, and T. R. Hayes, *Phys. Rev. A*, **41**, 3575 (1990)

111. R. H. McFarland and J. D. Kinney, *Phys. Rev.*, **137**, A1058 (1965)

112. M. Koparnski, Private communication (1991); D. Margreiter, H. Deutsch, and T. D. Märk, *Int. J. Mass Spectrom. Ion Proc.*, **139**, 127 (1994)

113. I. P. Zapesochnyi and I. S. Aleksakhin, *Zh. Eksp. Teor. Fiz.*, **55**, 76 (1968). English translation: *Sov. Phys. JETP*, **28**, 41 (1969)

114. G. O. Brink, *Phys. Rev.*, **127**, 1204 (1962)

115. K. L. Bell, H. B. Gilbody, J. G. Hughes, et al., *J. Phys. Chem. Ref. Data*, **12**, 891 (1983)

116. V. A. Astapenko, V. M. Buimistrov, and Iu. A. Krotov, *Zh. Eksp. Teor. Fiz.*, **93**, 825 (1987). English translation: *Sov. Phys. JETP*, **66**, 470 (1987)

117. K. Ishii and S. Morita, *Phys. Rev. A*, **30**, 2278 (1984)

118. Byung-Ho Choi, *Phys. Rev. A*, **4**, 1002 (1971)

119. O. I. Obolenskii, *Polyarizatsionnoe Tormoznoe Izluchenie Elektronov Vnutrennikh Obolochek Atomov* (Polarization Bremsstrahlung of Atomic Inner Shell Electrons) [in Russian], PhD Thesis, FTI, St-Petersburg (2000)

120. E. T. Verkhovtseva, E. V. Gnatchenko, and P. S. Pogrebnyak, *J. Phys. B: At. Mol. Phys.*, **16**, L613 (1983)

121. M. Ya. Amus'ya, T. M. Zimkina, and M. Yu. Kuchiev, *Zh. Tekh. Fiz.*, **52**, 1424 (1982). English translation: *Sov. Phys. Tech. Phys.*, **27**, 866 (1982)

122. T. M. Zimkina, A. S. Shulakov, and A. P. Braiko, *Fiz. Tverd. Tela*, **23**, 2006 (1981)

123. T. M. Zimkina, I. I. Lyakhovskaya, A. S. Shulakov, et al., *Fiz. Tverd. Tela*, **25**, 26 (1983)

124. E. T. Verkhovtseva, E. V. Gnatchenko, B. A. Zon, et al., *Zh. Eksp. Teor. Fiz.*, **98**, 797 (1990). English translation: *Sov. Phys. JETP*, **71**, 443 (1990)

125. A. A. Tkachenko, E. V. Gnatchenko, and E. T. Verkhovtseva, *Opt. Spektrosk.*, **78**, 208 (1995). English translation: *Opt. Spectrosc.*, **78**, 183 (1995)

126. H. Bethe and W. Heitler, *Proc. R. Soc. London Ser. A*, **146**, 83 (1934)

127. S. V. Blazhevich, G. L. Bochek, V. B. Gavrikov, et al., *Pis'ma Zh. Eksp. Teor. Fiz.*, **59**, 498 (1994). English translation: *JETP Lett.*, **59**, 524 (1994)

128. S. V. Blazhevich, A. S. Chepurnov, V. K. Grishin, et al., *Phys. Lett. A*, **211**, 309 (1996)

129. S. Blazhevich, A. Chepurnov, V. Grishin, et al., *Phys. Lett. A*, **254**, 230 (1999)

130. S. V. Blazhevich, V. K. Grishin, V. S. Ishkhanov, et al., *Izv. Vyssh. Uchebn. Zaved., Fiz.*, **44**, No. 3, 66 (2001). English translation: *Russ. Phys. J.*, **44**, 276 (2001)

131. V. L. Kleiner, N. N. Nasonov, and A. G. Safronov *Phys. Status Solidi B*, **181**, 223 (1994)

132. V. I. Iveronova and G. P. Revkevich, *Teoriya Rasseyaniya Rentgenovskikh Luchei* (Theory of X-Ray Scattering) [in Russian], Nauka, Moscow (1972)

133. V. M. Buimistrov, Yu. A. Krotov, and L. I. Trakhtenberg, *Zh. Eksp. Teor. Fiz.*, **79**, 808 (1980). English translation: *Sov. Phys. JETP*, **52**, 406 (1980)

134. K. Ishii, S. Morita, and H. Tawara, *Phys. Rev. A*, **13**, 131 (1976)

135. H. W. Schnopper, J. P. Delvaille, K. Kalata, et al., *Phys. Lett. A*, **47**, 61 (1974)

136. F. W. Saris, W. F. van der Weg, H. Tawara, and R. Laubert, *Phys. Rev. Lett.*, **28**, 717 (1972)

137. P. Kienle, M. Kleber, B. Povh, et al., *Phys. Rev. Lett.*, **31**, 1099 (1973)

138. D. H. Jakubassa and M. Kleber, *Z. Phys. A*, **273**, 29 (1975)

139. A. Weingartshofer, J. Holmes, G. Caudle, et al., *Phys. Rev. Lett.*, **39**, 269 (1977)

140. N. M. Kroll and K. M. Watson, *Phys. Rev. A*, **8**, 804 (1973)

141. E. L. Beilin and B. A. Zon, *J. Phys. B: At. Mol. Phys.*, **16**, L159 (1983)

142. B. Wallbank and J. K. Holmes, *Phys. Rev. A*, **48**, R2515 (1993)

143. S. Varró and F. Ehlotzky, *Phys. Lett. A*, **203**, 203 (1995)

144. L. Frommhold, *Collision-induced Absorption in Gases*, Cambridge Univ. Press, Cambridge (1993)

145. J. Jannssen, *C. R. Acad. Sci.*, **101**, 649 (1885)

146. M. Ya. Amus'ya, M. Yu. Kuchiev, and A. V. Solov'ev, *Pis'ma Zh. Eksp. Teor. Fiz.*, **10**, 1025 (1984). English translation: *Sov. Tech. Phys. Lett.*, **10**, 431 (1984)

147. G. C. Tabisz, E. J. Allin, and H. L. Welsh, *Can. J. Phys.*, **47**, 2859 (1969)

148. B. M. Smirnov, *Asimptoticheskie Metody v Teorii Atomnykh Stolknovenii* (Asymptotic Methods in Theory of Atomic Collisions) [in Russian], Atomizdat, Moscow (1973)

149. D. R. Bosomworth and H. P. Gush, *Can. J. Phys.*, **43**, 751 (1965)

150. U. Buontempo, S. Cunsolo, P. Dore, and P. Maselli, *J. Chem. Phys.*, **66**, 1278 (1977)

151. G. Herzberg, *Astrophys. J.*, **115**, 337 (1952)

152. V. M. Buimistrov, *Phys. Lett. A*, **30**, 136 (1969)

153. N. J. Mason, *Rep. Prog. Phys.*, **56**, 1275 (1993)

154. I. L. Beigman and B. N. Chichkov, *JETP Lett.*, **46**, 395 (1987)

155. R. J. Liefeld, A. F. Burr, and M. B. Chamberlain, *Phys. Rev. A*, **9**, 316 (1974)

156. M. B. Chamberlain, A. F. Burr, and R. J. Liefeld, *Phys. Rev. A*, **9**, 663 (1974)

157. G. Wendin and K. Nuroh, *Phys. Rev. Lett.*, **39**, 48 (1977)

158. K. Nuroh, in: *6th Intern. Conf. VUV Rad. Phys., Virginia, USA*, Ext. Abstracts, Vol. 1 (1980), p. 1

159. V. P. Romanikhin, *Zh. Eksp. Teor. Fiz.*, **121**, 286 (2002). English translation: *JETP*, **94**, 239 (2002)

160. V. A. Astapenko, L. A. Bureyeva, and V. S. Lisitsa, *Fiz. Plazmy*, **28**, 337 (2002). English translation: *Plasma Phys. Rep.*, **28**, 303 (2002)

161. V. A. Astapenko, *Priblizhennye Metody v Teorii Vzaimodeistviya Fotonov s Atomami* (Approximations in the Theory of Photons and Atoms Interaction) [in Russian], Izd. MFTI, Moscow (2002)

162. J. P. Connerade and A. V. Solov'yov, *J. Phys. B: At. Mol. Opt. Phys.*, **29**, 3529 (1996)

163. B. M. Smirnov, *Clusters and Small Particles in Gases and Plasmas*, Springer, New York (2000)

164. C. Brechignac and J. P. Connerade, *J. Phys. B: At. Mol. Opt. Phys.*, **27**, 3795 (1994)

165. L. Frommhold, *Phys. Rev. E*, **58**, 1899 (1998)

166. D. Hammer and L. Frommhold, *Phys. Rev. A*, **64**, 024705 (2001)

ASYMPTOTIC THEORY OF CHARGE EXCHANGE AND MOBILITY PROCESSES FOR ATOMIC IONS

B. M. Smirnov

1. Introduction

This paper is devoted to the asymptotic theory of the resonant charge exchange process in slow collisions and the theory of atomic ion kinetics in gases within an external electric field. Note the peculiarities of the asymptotic theory version under consideration. First, in contrast to models, the asymptotic theory uses a strict expansion of the cross section in power series of a small parameter that is inversely proportional to the typical distances between colliding particles at which the electron transfer proceeds. This allows one to estimate the accuracy of results under certain conditions. Second, we consider the simplest version of the asymptotic theory by taking into account two lower-order expansion terms. This allows us to express the cross section of resonant charge exchange through asymptotic parameters of a valence electron inside an isolated atom when the electron is located far from the atomic core. Third, if the electron transfer takes place between degenerated states of an atom and its ion, we find the mean cross section averaged over the electron-degenerated initial states of the colliding ion and atom, and this mean cross section pertains to a certain scheme of coupling of the ion and atom electron momenta and the rotational momentum of colliding atomic particles in the course of their collision. In this manner we find the mean cross section for a certain scheme of summation of electron momenta by averaging this cross section over the initial conditions. On this basis one can then estimate the accuracy of this mean cross section.

Basically we use a simple version of the asymptotic theory for the resonant charge exchange process in slow collisions, which first considers the asymptotic behavior of valence electrons in an isolated atom. We then split electron levels for an atom and its ion at large separations, and thus find the cross section of the electron transfer process, expressed through asymptotic parameters of a valence electron in an isolated atom, through atom and ion quantum numbers and through the collision velocity. The trade-off for simplifications is in the resulting accuracy which depends on the complexity of colliding particles and can not be improved within the framework of this version. This accuracy decreases with the complexity of colliding particles and in many cases is restricted by our knowledge of the asymptotic parameters of isolated atoms and ions. Nevertheless the accuracy of the asymptotic theory version under consideration is usually higher than that for experimental cross sections of the resonant charge exchange process, with a possible exception of cross sections obtained from mobilities of inert gas ions in parent gases. For this reason, experimental data for the cross sections of resonant charge exchange occupy a small part of this paper, the primary objective of which is to formulate the asymptotic theory of resonant charge exchange in the simplest form.

The accuracy of the cross sections obtained from the asymptotic theory is higher when the collision velocity is lower. Therefore the best application of this theory relates to kinetics of atomic ions in a parent atomic gas within an external electric field. The second part of this paper is devoted to the mobility and diffusion of atomic ions in parent and foreign gases. As the cross section of resonant charge exchange exceeds the cross section of ion – atom elastic scattering at high collision energies, the mobility of atomic ions in parent gases is expressed through the resonant charge exchange cross section. In considering some aspects of the kinetics of atomic ions in gases in an electric field, we follow specific methodical approaches that simplify the analysis and can be useful additions to general concepts and data for the ion mobility in gases [1 – 8].

Accordingly, the aim of this paper is to formulate the asymptotic theory of the charge exchange process in slow collisions involving atomic ions and to consider some aspects of the kinetic theory for the drift of atomic ions in gases with account for the resonant electron transfer cross sections.

2. Asymptotic theory of the interaction of atomic ions with parent atoms at large separations

2.1. Character of the resonant charge exchange process

The resonant charge exchange process proceeds according to the scheme:

$$A^+ + A \to A + A^+, \tag{2.1}$$

As a result of this process, a valent electron transfers from the field of one atomic core to another. In slow collisions, when the collision velocity is small compared to the atomic velocity, the rate of this process is expressed through parameters of electron terms of a quasimolecule constituted from the colliding particles. This process was first analyzed by Massey et al [9, 10] on the basis of the phase theory of collisional processes. Sena [11 – 13] used the classical character of motion of colliding particles. This enabled him to ascertain the physical nature of the process and to find the dependence of the cross section on collisional parameters. In particular, the resonant charge exchange cross section σ_{res} depends on the relative velocity of collision v as [11, 14]

$$\sigma_{res} = \frac{\pi}{2\gamma^2} \ln^2 \frac{v_0}{v}. \tag{2.2}$$

Here $\gamma = \sqrt{2I}$, I is the ionization potential of an atom A, and parameter $v_0 \gg 1$. Since formula (2.2) may be rewritten in the form

$$\sigma_{res} = \frac{\pi}{2\gamma^2}(\gamma R_0)^2,$$

below we shall apply the asymptotic theory of the resonant charge exchange process which uses a small parameter $1/\gamma R_0$. We determine the cross sections within the framework of the asymptotic theory $\gamma R_0 \gg 1$ under real conditions of ion – atom collisions.

Analyzing the resonant charge exchange process, we focus on the theoretical evaluation of the cross sections for three reasons. First, the accuracy of the asymptotic theory is better than the experimental results (possibly, excluding the case of inert gases, if the resonant charge exchange cross sections is derived from the mobility of ions in parent gases). Second, experimental data is restricted to a part of elements, and the collision energies are limited for some elements. Third, in many cases we can not estimate the accuracy of experimental data. Hence the experimental studies provide a restricted contribution to the data about resonant charge transfer. Below we will use the experimental data only for demonstrating advantages of this theory.

The cross section of a slow process can be expressed through parameters of electron terms responsible for this process. The eigenstates of the quasimolecule A_2^+ are divided in odd and even states in accordance with the wave function properties of these states, allowing conservation or change of the sign as a result of reflection of electrons with respect to the symmetry plane that is perpendicular to the molecular axis and bisects it. If initially an atom A and ion A^+ have only one electron state, there are only one even and one odd quasimolecule state, with the wave functions $\psi_g(\mathbf{r}, R), \psi_u(\mathbf{r}, R)$, and energies $\varepsilon_g(R), \varepsilon_u(R)$. Here $\mathbf{r}$ defines electron coordinates, R is the distance between the nuclei. At large separations we have

$$\psi_g = \frac{1}{\sqrt{2}} (\psi_1 + \psi_2), \quad \psi_u = \frac{1}{\sqrt{2}} (\psi_1 - \psi_2), \qquad (2.3)$$

where the wave functions ψ_1, ψ_2 correspond to electron locations in the fields of the first or second ion correspondingly (see Fig. 1).

Assuming the absence of inelastic transitions, one can construct a molecular wave function Ψ provided that before the collision $t \to -\infty$ the electron is bound with the first atomic core ($\Psi(\mathbf{r}, \mathbf{R}, t = -\infty) = \psi_1(\mathbf{r})$). Since two quasimolecule states are developed independently, we have

$$\Psi(\mathbf{r}, \mathbf{R}, t) = \frac{1}{\sqrt{2}} \psi_g(\mathbf{r}, \mathbf{R}) \exp\left[-i \int\limits_{-\infty}^{t} \varepsilon_g(t') \, dt' \right]$$

$$+ \frac{1}{\sqrt{2}} \psi_u(\mathbf{r}, \mathbf{R}) \exp\left[-i \int\limits_{-\infty}^{t} \varepsilon_u(t') \, dt' \right]. \qquad (2.4)$$

Here the relative motion of nuclei $R(t)$ is introduced which for free motion has the form $R = \sqrt{v^2 t^2 + \rho^2}$, where v is the collision ve-

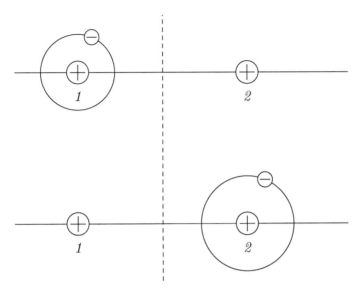

Fig. 1. Electron in the field of two identical centers. Reflection in the symmetry plane corresponds to the transformation $\psi_1 \to \psi_2$, $\psi_2 \to \psi_1$. This yields the eigenfunctions of the system, so that the symmetric electron wave function which retains its sign under reflection is $\psi_g = \frac{1}{\sqrt{2}}(\psi_1 + \psi_2)$ and the antisymmetric wave function is $\psi_u = \frac{1}{\sqrt{2}}(\psi_1 - \psi_2)$.

locity and ρ is the collision impact parameter. From this we find the probability P_{res} of the charge exchange process and its cross section [15]:

$$P_{\text{res}} = \sin^2 \zeta(\rho)\,, \quad \zeta(\rho) = \int\limits_{-\infty}^{\infty} \frac{\Delta(R)}{2}\, dt\,,$$

$$\Delta(R) = \varepsilon_g - \varepsilon_u \quad \sigma_{\text{res}} = \int\limits_{0}^{\infty} 2\pi\rho\, d\rho \sin^2 \zeta(\rho)\,.$$

(2.5)

Formula (2.5) expresses the parameters of the charge exchange process through electron terms $\varepsilon_g(R), \varepsilon_u(R)$ of the quasimolecule consisting of colliding particles. This connection was established by Firsov [15] and Demkov [14].

One more peculiarity of this process in slow collisions is the large cross section compared to the typical atomic value. This allows us to construct an asymptotic theory [16 – 18] which represents the cross section as a result of expansion over a small parameter $1/(\sqrt{\sigma}\gamma)$. In this case, restricting ourselves to using two terms in the expansion in series of the small parameter, we have for the charge exchange cross section [16 – 18]:

$$\sigma_{\mathrm{res}} = \frac{\pi R_0^2}{2}, \quad \text{where} \quad \zeta(R_0) = \frac{e^{-C}}{2} = 0.28\,. \tag{2.6}$$

Here $C = 0.577$ is the Euler constant.

Thus, within the asymptotic theory of the resonant charge exchange process, we suppose the electron transition to occur at large distances between nuclei compared to the orbit size of the transferring electron. Then we use the asymptotic expression of the exchange interaction potential of the ion and atom $\Delta(R) = \varepsilon_g - \varepsilon_u$ and this value is expressed in turn through asymptotic parameters of the atomic wave function at large distances of the electron from its atomic core. In this version of the asymptotic theory we do not use the electron distribution inside the atom, and information about the electron behavior inside the atom is included in the theory indirectly through the asymptotic coefficient of the valence electron.

2.2.   Ion – atom exchange interaction potential

We divide the problem of determination of the cross section of resonant charge exchange in to two steps. First we determine the exchange interaction potential of an ion with the parent atom, and next we connect the cross section with the exchange interaction potential. Below we solve this problem in a general form for the atoms and ions in the ground states or in lower excited states if atoms and ions have the same electron shell as in the ground state. We begin by solving for the exchange interaction in the simplest case if an s-electron is found in the field of two structureless ions.

The exchange interaction potential of atomic particles is determined by overlapping of electron wave functions belonging to different atomic centers. Below we determine the exchange interaction potential of an ion with the parent atom connected with transition of a valent electron from the field of one ion to the field of another. The character of this interaction due to overlapping of the electron wave functions is given in Fig. 2. We first consider the case when the valent electron is found in an s-state so that the considered system has two states that can be composed from states related to location of the electron in the field of the first and second ion (see Fig. 1).

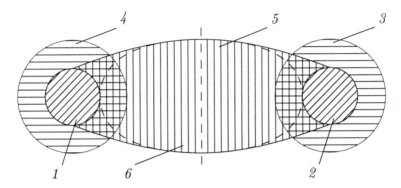

Fig. 2. Electron regions which determine the exchange interaction potential for an ion and parent atom at large distances between nuclei. *1,2* are internal regions of atoms where the electrons are located; *3,4* are regions where the asymptotic expressions for the atomic wave functions are valid; *5* is the region where the quasiclassical approach is valid for valent electrons (it is restricted by the dotted line); *6* is the region which mostly contributes to the exchange interaction potential of these atoms. The volume of region *6* is of the order of R^2, where R is the distance between nuclei, and regions *1,2* occupy a volume of the order of the atomic value. Using asymptotic data for atomic wave functions one can evaluate the exchange interaction potential to an accuracy of the order of $1/R^2$.

Let us denote by ψ_1 the electron wave function that is centered on the first nucleus, and by ψ_2 the wave function centered on the second nucleus. The electron Hamiltonian has the form

$$\hat{H} = -\frac{1}{2}\Delta + V(r_1) + V(r_2) + \frac{1}{R}. \qquad (2.7)$$

Here R is the distance between atomic cores, r_1, r_2 are the distances of the electron from the corresponding nucleus, $V(r)$ is the electron – ion interaction potential which takes the Coulomb form $V(r) = -1/r$ far from the ion. The symmetry of the problem under consideration implies that the symmetry plane is perpendicular to the line connecting nuclei and bisects it. The electron reflection with respect to this plane conserves the electron Hamiltonian. Hence, the electron eigenstates can be divided into even and odd, depending on the property of their wave functions to conserve or change their sign as a result of reflection with respect to the symmetry plane. Evidently, at large separations these wave functions take the form of the following compositions of ψ_1 and ψ_2 which correspond to location of the electron in the field of the corresponding atomic core:

$$\psi_g = \frac{1}{\sqrt{2}} (\psi_1 + \psi_2), \quad \psi_u = \frac{1}{\sqrt{2}} (\psi_1 - \psi_2). \qquad (2.8)$$

These wave functions of interacting ion and atom satisfy the Schrödinger equations

$$\hat{H}\psi_g = \varepsilon_g \psi_g, \quad \hat{H}\psi_u = \varepsilon_u \psi_u \qquad (2.9)$$

where $\varepsilon_g(R), \varepsilon_u(R)$ are the eigenvalues for the energies of these states. We define the exchange interaction potential in this case as

$$\Delta(R) = \varepsilon_g(R) - \varepsilon_u(R). \qquad (2.10)$$

In order to determine this value at large distances between nuclei, we use the following method [15]. Let us multiply the first equation (2.9) by ψ_u^*, the second equation by ψ_g^*, take the difference of the obtained equations and integrate the result over the volume Ω which is a half-space restricted by the symmetry plane. Since the separation between the nuclei is large, the wave function ψ_2 is zero inside this volume and the wave function ψ_1 is zero outside this volume. Hence $\int_\Omega \psi_u^* \psi_g \, d\mathbf{r} = 1/2$, and the relation obtained has the form

$$\frac{\varepsilon_g(R) - \varepsilon_u(R)}{2} = \frac{1}{2} \int_\Omega (\psi_u \Delta \psi_g - \psi_g \Delta \psi_u) \, d\mathbf{r}$$

$$= \frac{1}{2} \int_S (\psi_2 \frac{\partial}{\partial z} \psi_1 - \psi_1 \frac{\partial}{\partial z} \psi_2) \, d\mathbf{s},$$

where S is the symmetry plane which restricts the integration region. We use real wave functions in relations (2.8) and the z axis connects the nuclei. Take the origin of the reference frame in the center of the line connecting nuclei. Since the electron is found in the s-state in the field of each atomic core, its wave functions in this coordinate system can be represented in the form

$$\psi_1 = \psi\left(\sqrt{\left(z + \frac{R}{2}\right)^2 + \rho^2}\right),$$

$$\psi_2 = \psi\left(\sqrt{\left(z - \frac{R}{2}\right)^2 + \rho^2}\right),$$

(2.11)

where ρ is the distance from the axis in the perpendicular direction to it. Since $d\mathbf{s} = 2\pi\rho\,d\rho$, we obtain from the above relation [15]

$$
\varepsilon_g(R) - \varepsilon_u(R) = \int_0^\infty 2\pi\rho\,d\rho \left[\psi\left(\sqrt{\left(z - \frac{R}{2}\right)^2 + \rho^2}\right)\right.
$$

$$
\times \frac{\partial}{\partial z}\psi\left(\sqrt{\left(z + \frac{R}{2}\right)^2 + \rho^2}\right)
$$

$$
\left. - \psi\left(\sqrt{\left(z + \frac{R}{2}\right)^2 + \rho^2}\right)\frac{\partial}{\partial z}\psi\left(\sqrt{\left(z - \frac{R}{2}\right)^2 + \rho^2}\right)\right]_{z=0}
$$

$$
= R\int_0^\infty d\rho^2 \frac{\partial}{\partial \rho^2}\psi^2\left(\sqrt{\frac{R^2}{4} + \rho^2}\right) = R\,\psi^2\left(\frac{R}{2}\right). \quad (2.12)
$$

In deriving of this formula we employed the obvious relation

$$
\frac{\partial}{\partial z}\left[\psi\left(\sqrt{\left(z - \frac{R}{2}\right)^2 + \rho^2}\right)\right]_{z=0} = R\frac{\partial}{\partial \rho^2}\psi\left(\sqrt{\frac{R^2}{4} + \rho^2}\right).
$$

Now let us connect the molecular wave function $\psi(r)$ of the s-electron with the atomic wave function ψ_{at} which at large electron distances from the atomic core is determined by the Schrödinger equation

$$
-\frac{1}{2}\frac{\partial^2}{\partial r^2}(r\psi_{at}) - \frac{1}{r}\psi_{at} = -\frac{\gamma^2}{2}\psi_{at}
$$

where $\gamma^2/2$ is the electron binding energy. The solution of this equation is given by formula

$$\psi_{\text{at}}(r) = A r^{1/\gamma - 1} e^{-r\gamma}.$$ (2.13)

Take the molecular wave function in the form $\psi(r) = \chi(\mathbf{r})\psi_{\text{at}}(r)$ and compare the Schrödinger equations for molecular and atomic wave functions near the axis and far from nuclei where one can use the asymptotic form of the interaction potential $V(r) = -1/r$ in formula (2.7) for the electron Hamiltonian. So, neglecting the second derivative of χ near the axis, we have from the Schrödinger equation for ψ:

$$\gamma \frac{\partial \chi}{\partial r_1} + \left(\frac{1}{R} - \frac{1}{r_2} \right) \chi = 0.$$

Solving this equation, we connect the molecular wave function of the s-electron near the axis with the atomic wave function that allows us to express the exchange ion–atom interaction potential through asymptotic parameters of the valent s-electron in the atom [17]:

$$\Delta = A^2 R^{\frac{2}{\gamma} - 1} e^{-R\gamma - \frac{1}{\gamma}}.$$ (2.14)

In particular, this formula yields the exchange interaction potential of the proton and hydrogen atom in the ground state ($\gamma = 1$, $A = 2$) [19, 20]

$$\Delta = \frac{4}{e} R e^{-R}.$$

Formula (2.14) is the asymptotic expression for the exchange interaction potential of a one-electron atom with a valent s-electron and its atomic core. The criterion of validity of this formula reads:

$$R\gamma \gg 1, \quad R\gamma^2 \gg 1.$$ (2.15)

Generalizing formula (2.14) to the interaction of an one-electron atom with the parent ion, if the electron angular momentum is l and its projection onto the molecular axis is μ, we represent the electron wave function in the form

$$\psi(\mathbf{r}) = Y_{l\mu}(\theta, \varphi) \Phi(r),$$

where r, θ, φ are the spherical electron coordinates (provided that its center coincides with the corresponding nuclei) and the z axis is directed along the molecular axis. To determine the exchange interaction potential, we proceed analogously as in deriving formula (2.14) for the s-electron and make changes in integration with respect to $d\rho$. Then we have

$$\Delta \sim \int_0^\infty |Y_{l\mu}(\theta, \varphi)|^2 \, \Phi^2(r) \rho \, d\rho$$

where $r = \sqrt{R^2/4 + \rho^2}$ is the distance from each nucleus to the electron located in the symmetry plane. Since $\Phi(r) \sim e^{-\gamma r}$ the integral converges at small ρ ($\rho \sim \sqrt{R/\gamma} \ll R$) (see also Fig. 2). Then $\Phi(r) = \Phi(R/2)e^{-\gamma \rho^2/R}$. This corresponds to small angles $\theta = 2\rho/r$, and since $Y_{l\mu}(\theta, \varphi) \sim \theta^\mu$ for $\theta \ll 1$, we have

$$\Delta_\mu = \Delta_0 \int_0^\infty e^{-2\gamma \rho^2/R} |Y_{l\mu}(0,0)|^2 \left| \frac{2\rho}{R} \right|^{2\mu} \cdot 4\gamma \rho \, d\rho/R,$$

where Δ_0 is the exchange interaction potential given by formula (2.14) in the case of zero angular momentum of the valent electron with the same radial wave function. Since the exchange interaction potential does not depend on the sign of μ, we assume the momentum projection to be positive. Thus, we find for the exchange interaction potential of a one-electron atom with the parent ion [18, 21, 22]

$$\Delta_\mu = A^2 R^{\frac{2}{\gamma} - 1 - \mu} e^{-R\gamma - \frac{1}{\gamma}} \cdot \frac{(2l+1)(l+\mu)!}{(l-\mu)!\mu!(2\gamma)^\mu}. \tag{2.16}$$

In the case of structureless cores and nonzero electron momentum the ion–atom exchange interaction potential is characterized by the electron momentum projection m onto the molecular axis and is given by [18, 21–23]

$$\Delta_{l\mu} = \Delta_0 \frac{(2l+1)(l+|m|)!}{(l-|m|)! \, |m|!(R\gamma)^{|m|}} \tag{2.17}$$

where l is the electron orbital momentum and Δ_0 is the exchange interaction potential of the s-electron with the same asymptotic radial wave function (2.13) of the transferring electron. Formula (2.17) describes the ion–atom exchange interaction potential if the atom has one valent electron at large separations according to criterion (2.15). This interaction potential is determined by the overlapping of electron wave functions corresponding to location of the electron in the field of the first and second cores (see Fig. 2). According to the criterion (2.15), this formula is not suitable for highly excited atoms and relates to the ground and lowest excited atomic states.

2.3. Ion–atom exchange interaction for light atoms

We now generalize formula (2.17) for a light atom and ion with noncompleted electron shells when spin-orbit splitting of atom and ion levels is small and neglect relativistic interactions in them. This corresponds to the LS-coupling scheme for the atom. At large separations the quantum numbers of the molecular ion are the atomic quantum numbers LSM_LM_S (the orbital momentum, spin and their projections onto the molecular axis), and the same quantum numbers of the ion are $lsmm_s$. We sum up the electron orbital momentum and spin $l_e, \frac{1}{2}$ and these momenta of the atomic core ls into the atom momenta LS, and then the atom spin S and the spin of another atom core s are summed into the total spin I of the molecular ion. Then the atomic wave function is expressed through parameters of the core and valent electron by means of the genealogical or Racah coefficients [24–26], and the ion–atom exchange interaction potential has the form [18, 22]

$$\Delta(l_e\mu, lms, LM_LS) = \frac{\bar{I}+\frac{1}{2}}{2s+1}\, n(G_{ls}^{LS})^2 \begin{bmatrix} l_e & l & L \\ \mu & m & m+\mu \end{bmatrix}$$
$$\times \begin{bmatrix} l_e & l & L \\ \mu & M_L-\mu & M_L \end{bmatrix} \Delta_{l_e\mu} \qquad (2.18)$$

where n is the number of identical valent electrons, G_{ls}^{LS} is the parentage or Racah coefficient [24–26], the square brackets are the Clebsh–Gordan coefficients responsible for summation of electron and ion orbital momenta into the atomic orbital momentum, and $\Delta_{l_e\mu}$ is the exchange interaction potential for one valent electron which is located in the field of the structureless cores. Note a weak dependence of the molecular ion energies on the total spin I of the molecular ion. Indeed the level splitting corresponding to different total spins of the molecular ion is determined by exchange of two electrons and varies at large separations R as $\exp(-2\gamma R)$. Therefore formula (2.18) contains the average spin of the molecular ion. Next, since the exchange interaction potential $\Delta_{l_e \mu}$ decreases with increasing μ as $R^{-\mu}$, we are restricted in formula (2.18) by the term with minimal value of μ. As a result, in the case of a valent p-electron we have [18, 21 – 23]

$$\Delta_{10}(R) = 3\Delta_0, \quad \Delta_{11}(R) = \frac{6}{R\gamma}\Delta_0 \qquad (2.19)$$

where Δ_0 is the exchange interaction potential for a valent s-electron with the same asymptotic radial wave function (2.13) that is given by formula (2.14).

Formula (2.18) allows one to construct the matrix of the exchange interaction potential of an ion and atom with valent p-electrons. Below we represent these matrices if the atom and ion are located in the ground electron states. One can check the identity of the transferring electron and a hole. For atoms of group III (one valent p-electron) and atoms of group VIII (one valent p-hole) of the periodic table of elements, when the ground states of the atom and ion are ^1S and ^2P, respectively, the exchange interaction potential of the interacting atom and ion according to formula (2.18) is given by the matrix

$$\Delta(M_L) = \begin{array}{|c|c|c|} \hline M_L = -1 & M_L = 0 & M_L = +1 \\ \hline \Delta_{11} & \Delta_{10} & \Delta_{11} \\ \hline \end{array} \qquad (2.20a)$$

where M_L is the orbital momentum projection for the atom (elements of group III) or ion (elements of group VIII). For elements of groups IV and VII of the periodic table, when the ground electron states of the atom and ion are ^3P and ^2P, respectively, the exchange interaction potential matrix according to formula (2.18) is

$$\Delta(m, M_L) = \frac{5}{3} \cdot$$

	$M_L = -1$	$M_L = 0$	$M_L = +1$
$m = -1$	Δ_{10}	Δ_{11}	Δ_{10}
$m = 0$	Δ_{11}	$2\Delta_{11}$	Δ_{11}
$m = 1$	Δ_{10}	Δ_{11}	Δ_{10}

(2.20b)

where m, M_L are the projections of the orbital ion and atom momenta. For elements of groups V and VI of the periodic table with the ground states of the atom and ion ^4S and ^3P, respectively, the exchange interaction potential matrix has the form

$$\Delta(m) = \frac{7}{3} \cdot$$

$m = -1$	$m = 0$	$m = 1$
Δ_{11}	Δ_{10}	Δ_{11}

(2.20c)

We take as a quantization axis the direction in which the projection of the electron momentum is zero and we denote by θ the angle between the quantization and molecular axes. By definition, the exchange interaction potential $\Delta(\theta)$ of an atom and its ion with valent p-electrons is equal to

$$\Delta(\theta) = \frac{1}{3} \sum_M \left| d^1_{M0}(\theta) \right|^2 \Delta_{1M} = \frac{4\pi}{3} \sum_M |Y_{1M}(\theta, \varphi)|^2 \Delta_{1M}$$

where $d^1_{M0}(\theta)$ is the Wigner function of rotation [27], and $Y_{1M}(\theta)$ is the spherical function. From this it follows that $4\pi |Y_{1M}(\theta)|^2$ is the probability to find a state with the momentum projection M at angles θ, φ with respect to the molecular axis. The spherical function satisfies to the normalization condition

$$\int_{-1}^{1} d\cos\theta \, |Y_{1M}(\theta)|^2 = \frac{1}{4\pi}$$

and $-1 \le \cos\theta \le 1$. Hence we obtain the exchange interaction potential of an atom and the parent ion in the case of groups III and VIII of the periodic table of elements [18] in the form:

$$\Delta(\theta) = \Delta_{10} \cos^2\theta + \Delta_{11} \sin^2\theta. \qquad (2.21a)$$

Matrix (2.20b) gives the ion–atom exchange interaction potential as a function of the angles between the quantization and molecular axes for elements of groups IV and VII of the periodic table [28]

$$\Delta(\theta) = \frac{5}{3} \left[\Delta_{10} \sin^2 \theta_1 \sin^2 \theta_2 + \Delta_{11} (\cos^2 \theta_1 + \cos^2 \theta_2) \right] \qquad (2.21b)$$

where θ_1, θ_2 are the angles between the molecular axis and the quantization axis for the atom and ion correspondingly, with the result that the electron momentum projection onto the quantization axis is zero. In the case of groups IV and VII of the periodic table the exchange interaction potential is similar to that for atoms of groups III and VIII and has the form

$$\Delta(\theta) = \frac{7}{3} \cdot \left(\Delta_{10} \cos^2 \theta + \Delta_{11} \sin^2 \theta \right) . \qquad (2.21c)$$

Though we are restricted by the ground states of an ion and the parent atom, this is a general scheme of construction of the ion–atom exchange interaction potential. Being averaged over the total quasimolecule spin I, the exchange interaction potential depends on the ion m and atom M_L angular momentum projections onto the molecular axis. This corresponds to the LS-coupling for atoms and ions, i.e. we neglect the spin-orbital interaction. Hence, the above expressions correspond to the following hierarchy of the interaction potentials

$$V_{\text{ex}} \gg U(R), \Delta(R), \qquad (2.22)$$

where V_{ex} is the typical exchange interaction potential for valent electrons inside the atom or ion, $U(R)$ is the long-range interaction potential between the atom and ion at large separations R, $\Delta(R)$ is the exchange interaction potential between the atom and ion. Within the framework of the LS coupling scheme for atoms and ions, we assume the excitation energies inside the electron shell to be relatively large, and this criterion is fulfilled for light atoms and ions. In the same manner one can construct the exchange interaction potential matrix for excited states within a given electron shell.

Because the exchange interaction potential is determined by the transition of one electron from a valent electron shell and since

Table 1. The states of an ion and the parent atom with valent p-electrons. A one-electron transition is forbidden between these states, and the exchange interaction potential of the ion and parent atom is zero.

Ion – electron configuration and state	Atom – electron configuration and state
$p^2(^1D)$	$p^3(^4S)$
$p^2(^1S)$	$p^3(^4S)$
$p^2(^1S)$	$p^3(^2D)$
$p^3(^4S)$	$p^4(^1D)$
$p^3(^4S)$	$p^4(^1S)$
$p^3(^2D)$	$p^4(^1S)$

a transferring electron carries a certain momentum and spin, additional selection rules hold for one-electron interaction. In the case of transition of a p-electron, the selection rules have the following form

$$|L - l| \leq 1, \quad |S - s| \leq 1/2. \tag{2.23}$$

These selection rules follow from the properties of the Clebsh – Gordan coefficients in formula (2.18). If these conditions are violated then the ion – atom exchange interaction potential is zero on the scale of one-electron interaction potentials. Table 1 [28] lists the states of atoms and their ions with valent p-electrons for which the ion – atom one-electron exchange interaction potential is zero.

2.4.   Ion – atom exchange interaction for heavy atoms

In the limit of interaction of atomic particles, if the relativistic interactions dominate, the jj-coupling becomes valid for an individual atomic particle. Therefore the quantum numbers of an interacting atom and ion include J, the total electron momentum M_j, its projection onto the molecular axis for the atom, and j, m_j, the same quantum numbers for the corresponding ion. At large separations these quantum numbers relate to the molecular ion consisting of the ion and the parent atom. We note that the total momentum J and its projection onto a given direction M_J are the quantum numbers of an individual atomic particle for both momentum couplings (LS and jj) that simplifies the analysis in a general case. Next, an accounting for relativistic effects reduces the atom symmetry. For this reason the ion–atom exchange interaction potential is expressed on the one hand through the one-electron exchange interaction potential in a simpler way, and on the other hand, the prohibition on some one-electron transitions becomes stronger in the presence of relativistic interactions because of a weaker mixing of states in this case. Table 2 [28, 29] contains parameters of electron shells for the ground electron states of atoms and ions having p electron shells. Note that in the case of jj-coupling, the analogy in transitions of a p-electron and a p-hole is lost because of a different sign of the spin-orbit interaction potential for the electron and the hole. Hence the ion–atom exchange interaction potential is different when the p-electron shells of an atom and its ion are replaced by the shells consisting of identical p-holes. Moreover, in the case of group VI elements of the periodic table, the one-electron ion–atom exchange interaction potential is zero if the atom and ion are found in the ground states. Note that for all the groups of the periodical table of elements with valent p-electrons, the ion–atom one-electron exchange interaction potential is not zero for light atoms and their ions in the ground states.

As follows from Table 2, the ion–atom exchange interaction potential is simpler in the presence of relativistic interactions because of a lower symmetry of the atomic particles in this case. In the case of LS-coupling for individual atomic particles we were restricted to the ground states of atomic particles because of a cumbrous problem, but the presence of relativistic effects simplifies this problem. To illustrate this fact, Table 3 gives the exchange interaction potential matrix for group V elements. The notations of ion and atom electron terms are indicated in Table 3 for LS- and jj-coupling. The values of the exchange interaction potentials are given if the jj momentum coupling holds true, and it is indicated in parentheses that this potential is zero (0) or it is not zero (+) for the LS-coupling. In particular, for the ground atom and ion states the exchange interaction

Table 2. The ground states of atoms with p-electron shells within the framework of LS- and jj-couplings, and the ion–atom exchange interaction potential (Δ) for the cases c and e of the Hund coupling.

Shell	J	LS-term	jj-shell	Δ
p	$1/2$	$^2P_{1/2}$	$[1/2]^1$	$\Delta_{1/2}$
p^2	0	$^3P_0$	$[1/2]^2$	$\Delta_{1/2}$
p^3	$3/2$	$^4S_{3/2}$	$[1/2]^2[3/2]^1$	$\Delta_{3/2}$
p^4	2	$^3P_2$	$[1/2]^1[3/2]^3$	0
p^5	$3/2$	$^2P_{3/2}$	$[1/2]^2[3/2]^3$	$\Delta_{1/2}$
p^6	0	$^1S_0$	$[1/2]^2[3/2]^4$	$\Delta_{3/2}$

potential occupies one cell in Table 3, while within the framework of the LS-coupling it is given by the matrix of formula (2.20c).

Note that for jj coupling the p-electron shell of an atom or ion is split into two independent subshells with $j = 1/2$ and $j = 3/2$. Hence, the difference in the number of electrons on these subshells for an interacting ion and atom can not exceed one. This is the criterion of one-electron transition instead of (2.23) for LS-coupling. If this criterion is not fulfilled, the one-electron ion–atom exchange interaction potential vanishes, and it is equal to $\Delta_{1/2}$ or $\Delta_{3/2}$ depending on the momentum of transferring electron (see Tables 2, 3).

We now focus on elements of groups III or VIII of the periodic table when one transferring p-electron (or p-hole) is located in the field of two structureless cores. If the spin-orbit splitting of electron levels is large compared to electrostatic ion–atom interaction, the quantum numbers of the molecular ion are jm_j — the total electron momentum and its projection onto the molecular axis. We have the following relations between the exchange interaction potential Δ_{jm_j} within the framework of jj-coupling for atoms and ions, and the exchange interaction potentials Δ_{1m} for the LS-coupling:

$$\Delta_{jm_j} = \sum_{\mu} \left[\begin{array}{ccc} \frac{1}{2} & 1 & j \\ \sigma & \mu & m_j \end{array} \right]^2 \Delta_{1\mu}.$$

Table 3. The exchange interaction potential for atoms of group V of the periodic system of elements with the atomic electron shell p^3 and their ions with the electron shell p^2 [29].

LS	jj	$^4S_{3/2}$ $\left[\left(\frac{1}{2}\right)^2\left(\frac{3}{2}\right)\right]_{3/2}$	$^2D_{3/2}$ $\left[\left(\frac{1}{2}\right)\left(\frac{3}{2}\right)^2\right]_{3/2}$	$^2D_{5/2}$ $\left[\left(\frac{1}{2}\right)\left(\frac{3}{2}\right)^2\right]_{5/2}$	$^2P_{1/2}$ $\left[\left(\frac{1}{2}\right)\left(\frac{3}{2}\right)^2\right]_{1/2}$	$^2P_{3/2}$ $\left[\left(\frac{3}{2}\right)^3\right]_{3/2}$
$^3P_0$	$\left[\left(\frac{1}{2}\right)^2\right]_0$	$\Delta_{3/2}(+)$	$0(+)$	$0(+)$	$0(+)$	$0(+)$
$^3P_1$	$\left[\left(\frac{1}{2}\right)\left(\frac{3}{2}\right)\right]_1$	$\Delta_{1/2}(+)$	$\Delta_{3/2}(+)$	$\Delta_{3/2}(+)$	$\Delta_{3/2}(-)$	$0(+)$
$^3P_2$	$\left[\left(\frac{1}{2}\right)\left(\frac{3}{2}\right)\right]_2$	$\Delta_{1/2}(+)$	$\Delta_{3/2}(+)$	$\Delta_{3/2}(+)$	$\Delta_{3/2}(-)$	$0(+)$
$^1D_2$	$\left[\left(\frac{3}{2}\right)^2\right]_2$	$0(0)$	$\Delta_{1/2}(+)$	$\Delta_{1/2}(+)$	$\Delta_{1/2}(+)$	$\Delta_{3/2}(+)$
$^1S_0$	$\left[\left(\frac{3}{2}\right)^2\right]_0$	$0(0)$	$\Delta_{1/2}(0)$	$\Delta_{1/2}(0)$	$\Delta_{1/2}(+)$	$\Delta_{3/2}(+)$

This relation follows from the relation between the electron wave functions for these states and yields the exchange interaction potentials Δ_{jm_j},

$$\Delta_{1/2,1/2} = \frac{1}{3}\Delta_{10} + \frac{2}{3}\Delta_{11}, \quad \Delta_{3/2,1/2} = \frac{2}{3}\Delta_{10} + \frac{1}{3}\Delta_{11},$$

$$\Delta_{3/2,3/2} = \Delta_{11}, \tag{2.24}$$

where $m_j = \sigma + \mu$ according to the properties of the Clebsh–Gordan coefficients. Here the values Δ_{10} and Δ_{11} are given by formulas (2.19).

By analogy with previous operations, if the molecular axis has angle θ with the quantization axis onto which the angular momentum projection is zero, the exchange interaction potentials is:

$$\Delta_{1/2} = \frac{1}{3}\Delta_{10} + \frac{2}{3}\Delta_{11},$$

$$\Delta_{3/2}(\theta) = \left(\frac{1}{6} + \frac{1}{2}\cos^2\theta\right)\Delta_{10} + \left(\frac{1}{3} + \frac{1}{2}\sin^2\theta\right)\Delta_{11}. \tag{2.25}$$

3. Asymptotic theory of resonant charge exchange process

3.1. Cross section of resonance charge exchange with transition of s-electron

We now evaluate the cross section of the resonant charge exchange process (2.1) when an s-electron goes from one atomic core to another. In this case we have on the basis of formulas (2.5) and (2.14) for free motion of nuclei $(R^2 = \rho^2 + v^2 t^2)$

$$\zeta_0(\rho) = \int\limits_{-\infty}^{\infty} \frac{\Delta(R)}{2}\,dt = \frac{1}{v}\sqrt{\frac{\pi\rho}{2\gamma}}\,\Delta(\rho)$$

$$= \frac{1}{v}\sqrt{\frac{\pi}{2\gamma}}\,A^2 e^{-1/\gamma}\rho^{2/\gamma-1/2}\exp(-\rho\gamma) \tag{3.1}$$

and the cross section according to formula (2.5) is [17, 18]

$$\sigma_{\text{res}} = \int 2\pi\rho\,d\rho\,\sin^2\zeta_0(\rho) = \frac{\pi R_0^2}{2} \tag{3.2a}$$

where

$$\zeta_0(\rho) = \frac{1}{v}\sqrt{\frac{\pi}{2\gamma}}\, A^2 e^{-1/\gamma} R_0^{2/\gamma - 1/2} \exp(-R_0\gamma) = 0.28\,. \qquad (3.2b)$$

In particular the velocity dependence (2) follows from this formula if the basic dependence $\zeta(R_0)$ is exponential.

In order to ascertain the accuracy of the asymptotic theory, we consider charge exchange of a proton on a hydrogen atom at a collision energy of 1 eV in the laboratory frame and analyze various versions of the asymptotic theory. In this case formula (3.2) takes the form

$$\sigma_{\text{res}} = \frac{\pi R_0^2}{2} \quad \text{where} \quad \zeta(R_0) = \frac{4}{v}\sqrt{\frac{\pi}{2}}\, R_0^{3/2} \exp(-1 - R_0) = 0.28\,.$$
$$(3.3a)$$

One can account for the next term in the expansion of the phase $\zeta(R_0)$ in the small parameter $1/R_0$. Then formula (3.2) has the form

$$\sigma_{\text{res}} = \frac{\pi R_0^2}{2}\,,$$
$$(3.3b)$$
$$\text{where } \zeta(R_0) = \frac{4}{v}\sqrt{\frac{\pi}{2}}\, R_0^{3/2}\left(1 + \frac{7}{8R_0}\right)\exp(-1 - R_0) = 0.28\,.$$

One can evaluate the exchange phase $\zeta(\rho)$ using the exchange interaction potential $\Delta(R)$ directly from formula (2.5). This gives for the charge exchange cross section

$$\sigma_{\text{res}} = \frac{\pi R_0^2}{2}\,, \quad \text{where} \quad \zeta(R_0) = \frac{4R_0^2}{ve}\left[K_0(R_0) + \frac{1}{R_0}K_1(R_0)\right] = 0.28\,.$$
$$(3.3c)$$

Finally, one can find the charge exchange cross section directly using formula (2.5)

$$\sigma_{\text{res}} = \int_0^\infty 2\pi\rho\, d\rho \sin^2 \zeta(\rho)\,, \qquad (3.3d)$$

where the charge exchange phase is given by formulas (3.3a), (3.3b), and (3.3c). The cross section in the hydrogen case at the collision energy of 1 eV in the laboratory frame calculated using the above

Table 4. The values of the parameter $R_0\gamma$ for resonant charge exchange accompanied by transition of s-electron at energy 1 eV in the laboratory frame.

Element	H	He	Li	Be	Na	Mg	K	Ca	Cu
$R_0\gamma$	10.5	10.5	13.6	12.7	14.9	13.8	15.7	14.7	14.3
Element	Zn	Rb	Sr	Ag	Cd	Cs	Ba	Au	Hg
$R_0\gamma$	14.0	16.3	15.4	14.5	14.4	16.8	16.0	14.5	14.5

formulas for values of the charge exchange cross section are in atomic units 172, 175, 175, correspondingly, if we use formulas (3.3a), (3.3b), and (3.3c), and 170, 173, 174 if we use formula (3.3d) with the above expressions for the charge exchange phase. The statistical treatment of this data gives [28] 173 ± 2 a.u. for the average cross section, i.e. the error in this case, which can be considered as the best accuracy of the asymptotic theory, is approximately 1%.

In reality, the accuracy of the asymptotic theory is determined by the small parameter $1/(R_0\gamma)$, and the above accuracy is of the order of $1/(R_0\gamma)^2$. Table 4 gives the values of the parameter $R_0\gamma$ for some cases of resonant charge exchange with s-electron transition at a collision energy of 1 eV in the laboratory frame. The values of the asymptotic coefficients A are taken from [18, 30, 31]. Clearly, the best accuracy of the asymptotic theory is less than 1%.

The accuracy of the asymptotic coefficient A is limited, especially if it is obtained from electron wave functions which are given as a sum of exponents [32, 33]. The error ΔA for this value affects the accuracy of the cross section. From formula (3.2) the relative accuracy of the cross section $\Delta\sigma$ is

$$\frac{\Delta\sigma_{\text{res}}}{\sigma_{\text{res}}} = \frac{4}{R_0\gamma} \cdot \frac{\Delta A}{A}. \tag{3.4}$$

Estimating the error in the asymptotic coefficient to be $\Delta A/A = 10\%$, we obtain the error in the cross section $3-4\%$ at a collision energy of 1 eV for the cases of Table 4 in accordance with formula (3.4). Thus the accuracy of the asymptotic coefficients is important for the asymptotic theory of the resonant charge exchange cross section, the real accuracy of which with the transferring s-electron lies between 1% and 5% at small collision energies.

According to Fig. 2, the contribution of the internal atomic regions to the overlapping integral is of order of $1/R^2$. Therefore, representing the charge exchange cross section as an expansion in series of the small parameter $1/R$, one can retain only two first terms of this expansion, the accounting for next terms being incorrect. Formula (3.2) takes into account two terms such an expansion. Hence this asymptotic theory is valid to a certain accuracy that can not be improved within the framework of the information used.

Note that the cross section of the resonant charge exchange process depends weakly on the collision velocity in accordance with formula (2.2). This dependence follows from the exponential dependence of the charge exchange phase $\zeta_0(\rho)$ on the collision impact parameter ρ $\zeta_0(\rho) \sim \exp(-\gamma\rho)$. In this case $\gamma R_0 = \ln(v_0/v)$, and we obtain

$$\frac{d\ln\sigma}{d\ln v} = \frac{2}{\gamma R_0} \, , \text{ or } \sigma(v) = \sigma(v_0) \cdot \left(\frac{v_0}{v}\right)^\delta \, , \quad \delta = \frac{2}{\gamma R_0} \, , \qquad (3.5)$$

and since $\gamma R_0 \gg 1$, we have $\delta \ll 1$.

3.2.   Cross section of resonant charge exchange with p-electron transition

The asymptotic theory is simple for transition of an s-electron when the exchange phase $\zeta(\rho)$ is given by formula (3.1). The resonant charge exchange cross section is determined by formula (3.2) which accounts for two leading terms in a series of expansion over a small parameter of the asymptotic theory. When a valent p-electron transfers from one atomic core to another during the collision, the processes of charge exchange and electron momentum rotation are entangled. One can partially separate these processes because charge exchange proceeds in a narrow range of separations where the molecular axis turns by a small angle of the order of $1/\sqrt{R_0\gamma}$. Indeed, a range of distances between nuclei ΔR, where the charge exchange phase ζ varies remarkably, is $\Delta R \sim 1/\gamma$, and this corresponds to the rotation angle $\vartheta \sim vt/R \sim 1/\sqrt{R\gamma} \ll 1$. Therefore, one can neglect the depolarization process in the course of the electron transition, but this decreases the accuracy of the asymptotic theory. Below we consider the transition of a p-electron in the resonant charge exchange process.

Separating in this way the depolarization of the colliding atom and ion from the charge exchange process, we average the cross section over directions of the molecular axis with respect to the quantization axis. Considering transition of a p-electron, we introduce an angle θ between the quantization axis for the electron momenta and the molecular axis of the colliding atom and ion. This value varies during the collision due to the molecular axis rotation. We denote by ϑ this angle at the distance of closest approach of the colliding particles, and the average resonant charge exchange cross section $\bar{\sigma}$ is equal to

$$\bar{\sigma}_{\text{res}} = \frac{1}{2} \int\limits_{-1}^{1} \sigma(\vartheta)\, d\cos\vartheta\,, \tag{3.6}$$

where $\sigma(\vartheta)$ is the charge cross section exchange at angle ϑ between the collision impact parameter and the quantization axis. Figure 3 shows the geometry of collision in the center-of-mass frame, when the configuration of the colliding particles is close to that at closest approach. The following expression relates to a current angle θ between the molecular and quantization axes and an angle ϑ between these axes at closest approach

$$\cos\theta = \cos\vartheta \cos\alpha + \sin\vartheta \sin\alpha \cos\varphi\,,$$

where α, φ are the polar angles of the molecular axis, so that $\sin\alpha = vt/R$, and v is the collision velocity, t is time, and R is the distance between the colliding particles.

A small parameter of the theory $1/\rho\gamma$ simplifies determination of the phase and cross section of this process. Formulas (2.21) give the expressions for the exchange interaction potentials of atoms and their ions with filled p-shells in neglecting the spin-orbit interaction. These expressions with accounting for the above relation between angles for molecular and quantization axes can be considered as an expansion of the exchange interaction potentials in power series of

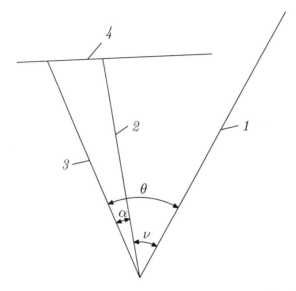

Fig. 3. Geometry of the collision process in the center-of-mass frame *1* — quantization axis; *2* — molecular axis at the distance of the closest approach; *3* — current molecular axis; *4* — trajectory of motion; θ, ϑ are the angles between the quantization and molecular axes for the current configuration of the colliding particles and at the distance of the closest approach, α, φ are polar angles of the current molecular axis with respect the one at the distance of the closest approach.

a small parameter $1/\rho\gamma$. As a result, the charge exchange phase in the case of atoms of groups III, V, VI, and VIII of the periodic table of elements is

$$\zeta(\rho, \vartheta, \varphi) = \zeta(\rho, 0) \left[\cos^2 \vartheta - \frac{1}{\gamma\rho} \cos^2 \vartheta + \frac{1}{\gamma\rho} \sin^2 \vartheta (2 + \cos^2 \varphi)\right].$$

$$(3.7)$$

This expression relates to large collision impact parameters, and $\zeta(\rho, 0)$ is the phase of the charge exchange process when a quantization axis has the same direction as the molecular axis at the distance of the closest approach. The value $\zeta(\rho, 0)$ can be expressed through the charge exchange phase ζ_0 which is given by formula (3.1) and relates to s-electron transition with the same asymptotic parameters γ, A. This relationship for the resonant charge exchange process involving atoms of groups III and VIII of the periodic table has the form

$$\zeta(\rho, 0) = 3\zeta_0(\rho), \tag{3.8a}$$

and for atoms of groups V and VI this relation is

$$\zeta(\rho, 0) = 7\zeta_0(\rho). \tag{3.8b}$$

Note that our analysis is valid for the ground state of the colliding atom and ion.

In the case of atoms of groups IV and VII of the periodic table the expression for the charge exchange phase at large impact collision parameters has the form

$$
\begin{aligned}
\zeta(\rho, \vartheta, \varphi) \;=\; 5\zeta_0(\rho)\bigg\{ & \sin^2 \vartheta_1 \sin^2 \vartheta_2 + \frac{1}{\gamma\rho}[2\cos^2 \vartheta_1 \\
& + 2\cos^2 \vartheta_2 + \sin^2 \vartheta_1 \cos^2 \vartheta_2 + \cos^2 \vartheta_1 \sin^2 \vartheta_2 \\
& - \sin^2 \vartheta_1 \sin^2 \vartheta_2 (\cos^2 \varphi_1 + \cos^2 \varphi_2) \\
& + \sin 2\vartheta_1 \sin 2\vartheta_2 \cos \varphi_1 \cos \varphi_2] \bigg\},
\end{aligned}
\tag{3.9}
$$

where ϑ_1, φ_1 and ϑ_2, φ_2 are the polar angles of the quantization axes of the atom and ion correspondingly with respect to the molecular axis at the distance of the closest approach.

We use formula (2.6) for the resonant charge exchange cross section at $\vartheta = 0$ and take into account that the angular dependence of the cross section is logarithmic according to formula (2.2), so that the average cross section of this process is close to that at zero angle. Hence, the average cross section can be determined as an expansion in power series of a small parameter of the theory. Indeed, since the basic dependence of the exchange phase $\zeta(\rho, \vartheta, \varphi)$ on a collision impact parameter ρ is exponential, $\zeta(\rho, \vartheta) \sim \exp(-\gamma\rho)$, we have in the case of atoms of groups III, V, VI, VIII of the periodic table

$$R_0(\vartheta, \varphi) = R_0(0) + \frac{1}{\gamma}\ln\frac{\zeta(\rho, \vartheta, \varphi)}{\zeta(\rho, 0)}. \tag{3.10}$$

We consider atoms and ions in the ground states, including the S-state, so that the angles ϑ, φ characterize the quantization axis direction for an atomic particle with nonzero momentum. Then formula (3.6) yields the average resonant charge exchange cross section

$$\bar{\sigma}_{\text{res}} = \frac{1}{4} \int\limits_0^1 \int\limits_0^{2\pi} \left[R_0(0) + \frac{1}{\gamma} \ln \frac{\zeta(R_0, \vartheta, \varphi)}{\zeta(R_0, 0)} \right]^2 d\cos\vartheta \, d\varphi \,. \tag{3.11}$$

In fact, formula (3.11) implies that the dependence of the exchange phase ζ on a collision impact parameter ρ has the form $\zeta \sim \exp(-\gamma\rho)$. This formula provides the basis for determination of the average resonant charge exchange cross section process when this process results from transition of a p-electron. This formula is valid for elements of groups III, V, VI, VIII of the periodic table when atoms and ions are in the ground states, and one of these states is the S-state, such that the charge exchange phase depends on two angles ϑ, φ. In the same manner one can find the charge exchange phase for elements of groups IV and VII which depend on four angles $\vartheta_1, \varphi_1, \vartheta_2, \varphi_2$.

Let us compare the resonant charge cross sections exchange for transition of s and p valent electrons when these electrons are characterized by the same asymptotic parameters γ and A. Assuming exponential dependence of the charge exchange phase $\zeta(\rho, \vartheta, \varphi)$ on the collision impact parameter ρ, $\zeta(\rho, \vartheta, \varphi) \sim \exp(-\gamma\rho)$, and neglecting the momentum rotation during the electron transition, we obtain the average cross section $\bar{\sigma}_{\text{res}}$ of the resonant charge exchange process

$$\bar{\sigma}_{\text{res}} = \sigma_0 \int\limits_0^1 d\cos\vartheta \int\limits_0^{2\pi} \frac{d\varphi}{2\pi} \left(1 + \frac{1}{\gamma R_0} \ln \frac{\zeta(\rho, \vartheta, \varphi)}{\zeta_0(\rho)} \right)^2 \,. \tag{3.12}$$

Here σ_0 is the resonant charge exchange cross section for a transferring s-electron with the same asymptotic parameters, $R_0 = \sqrt{2\sigma_0/\pi}$, and $\zeta_0(\rho)$ is the charge exchange phase for s-electron which is given by formula (3.1). Additional assumptions for evaluation of the charge exchange cross sections with a transferring p-electron decrease the accuracy of the asymptotic theory in this case.

Note that the above consideration pertains to the ground states of a colliding atom and ion. In the case of atoms and ions with valent p-electrons the cross section can vary significantly for different states. For example, for the process

$$O^+(^4S) + O \rightarrow O + O(^4S)$$

where the ion is found in the ground state, the cross section is zero for excited ^2D and ^2S atomic states of this p^3 valent shell. Because the statistical weight of the atom in the ground state is $3/5$ with respect to the total number of atomic states of this electron shell, the cross section of the process under consideration depends significantly on the method of preparing atoms and ions.

Table 5 contains the reduced cross sections depending on the parameter $R_0\gamma$. The value Σ_3 of Table 2 is $\bar{\sigma}/\sigma_0$ for elements of groups III and VIII of the periodic table, the value Σ_4 is $\bar{\sigma}/\sigma_0$ for elements of groups IV and VII of the periodic table, and the value Σ_5 is $\bar{\sigma}/\sigma_0$ for elements of groups V and VI of the periodic table. In addition, this Table contains the reduced cross sections Σ_{10} and Σ_{11} that correspond to the projections 0 and 1 of the orbital momentum onto the impact parameter direction corresponding to the case of one valence p-electron, i.e. for elements of groups III and VIII of the periodic table. The value $\bar{\Sigma}$ in Table 5 is $\bar{\Sigma} = \Sigma_{10}/3 + 2\Sigma_{11}/3$, and its comparison with the average cross section testifies the sensitivity of the cross section to different methods of averaging.

3.3. Resonant charge exchange for different cases of Hund coupling

In considering the resonant charge exchange as a process of slow collisions of atomic particles, we use the fact that the quantum numbers which describe this process depend on the character of momentum coupling for colliding atomic particles. In turn, this depends on energetic parameters of different degrees of freedom for colliding particles. Following the classical scheme [35, 36], we describe various relations between these energetic parameters as different cases of momentum coupling in the diatomic molecules as given in Table 6 [35, 34]. These are called the cases of Hund coupling and include

Table 5. Reduced cross sections of resonant charge exchange [29].

$R_0\gamma$	6	8	10	12	14	16
Σ_{10}	1.40	1.29	1.23	1.19	1.16	1.14
Σ_{11}	1.08	0.98	0.94	0.92	0.91	0.91
$\bar{\Sigma}$	1.19	1.08	1.04	1.01	0.99	0.95
Σ_3	1.17	1.09	1.05	1.03	1.02	1.01
Σ_3^d	1.16	1.08	1.04	1.02	1.01	1.00
Σ_4	1.50	1.32	1.23	1.18	1.14	1.12
Σ_5	1.44	1.29	1.22	1.17	1.14	1.11
$\Sigma_{1/2}$	1.18	1.10	1.07	1.05	1.04	1.03
$\Sigma_{3/2}$	1.18	1.10	1.06	1.04	1.03	1.02
$\Sigma_{3/2}^e$	1.16	1.09	1.06	1.04	1.03	1.02

different relations between energetic parameters of the colliding particles. The important energetic parameter of the quasimolecule consisting of colliding particles is the interaction potential V_e between the orbital angular momentum of electrons and the molecular axis. This includes the exchange interaction potential V_{ex} inside the atom and ion due to the Pauli exclusion principle, the splitting of molecular ion levels due to long-range interaction $U(R)$ and exchange interaction $\Delta(R)$ between the ion and atom. Within the framework of the Hund schemes, we compare this interaction potential with the relativistic interaction δ_f which is the sum of spin-orbit interactions of individual electrons and other relativistic interactions and the rotational energy $V_r = v\rho/R^2$ for free motion of colliding particles. In the case of colliding atomic particles, in contrast to that of a molecule, different types of Hund coupling can be realized in one classical trajectory of particles. We use a general Nikitin's scheme [23, 36 – 38] that connects the character of momentum coupling for colliding atomic particles in motion along one trajectory. The problem under consideration is simpler because the behavior of colliding particles is of interest on

Table 6. The cases of Hund coupling.

Hund coupling case	Relation
a	$V_e \gg \delta_f \gg V_r$
b	$V_e \gg V_r \gg \delta_f$
c	$\delta_f \gg V_e \gg V_r$
d	$V_r \gg V_e \gg \delta_f$
e	$V_r \gg \delta_f \gg V_e$
e'	$\delta_f \gg V_r \gg V_e$

the trajectory element where the electron transition proceeds. Only one type of the momentum coupling is realized on this part of the trajectory.

Above we considered the cases where relativistic interactions were negligible and orbital electron momentum projection onto the molecular or motionless axis was conserved during electron transition

$$\Delta(\rho), V(\rho) \gg \frac{v}{\rho^2}, \delta_f$$

corresponding to cases a and b of Hund coupling which were realized for light atomic particles. It is of interest to compare these results with those for the case d of Hund coupling which is given in Table 5 for elements of groups III and VIII of the periodic table. This cross section is denoted by Σ_3^d. Though the case d which corresponds to a motionless axis is not realized, the comparison shows that the cross sections are close in three cases a, b and d of Hund coupling.

We now consider the resonant charge exchange process within the framework of the case c of Hund coupling when

$$\delta_f \gg V_{ex} \gg V_r \tag{3.13}$$

according to the data of Table 6. This criterion leads to the jj-coupling in the atom and ion, which in turn corresponds to transition of one electron with a given total momentum j during resonant charge exchange. Below we analyze the character of resonant charge transfer for atoms and ions with valent p-electrons. In the case c of Hund coupling and transition of a p-electron or a p-hole between two completed cores, we have only one electron term, if the electron momentum is $1/2$. The exchange interaction potential for this fine state of the atom or ion is given by formula (2.25), and this leads to the following exchange phase

$$\zeta_{1/2}(\rho,\vartheta) = \zeta_0(\rho)\left(1 + \frac{4}{\rho\gamma}\right) , \qquad (3.14)$$

where $\zeta_0(\rho)$ is the charge exchange phase defined according to formula (3.1), so that $3\zeta_0(\rho)$ is the charge exchange phase for zero projection of the electron momentum onto the collision impact parameter in the case a of Hund coupling. In the case where the total electron momentum is $3/2$ the exchange phase within the framework of the case c of Hund coupling follows from formulas (2.25)

$$\zeta_{3/2}(\rho,\vartheta) = \zeta_0(\rho)\left[\frac{1}{2} + \frac{3}{2}\cos^2\vartheta \right.$$
$$\left. + \frac{1}{\rho\gamma}\left(\frac{1}{2} + \frac{9}{2}\sin^2\vartheta + \frac{3}{2}\sin^2\vartheta\cos^2\varphi\right)\right] . \quad (3.15)$$

Here $\zeta_0(\rho)$ is the charge exchange phase for the transition of an s-electron with the same asymptotic parameters A,γ which is defined according to formula (2.4), and ϑ,φ are the polar angles of the impact parameter direction with respect to the quantization axis. Table 2 contains the reduced cross sections $\Sigma_j = \overline{\sigma_j}/\sigma_s$, where the average cross section $\overline{\sigma_j}$ for a given total momentum is determined by formula (2.13). The difference between average cross sections for different total momenta is small in comparison with the accuracy of determination of the cross sections and is negligible. One can determine the cross sections for the case e of Hund coupling when due to a large rotational energy the momentum projection conserves onto the motionless axis for the state with $j = 3/2$, and $\Sigma^e_{3/2}$ in Table 2 is the reduced cross section of the resonant charge exchange for the state with $j = 3/2$ in the case e of Hund coupling. According to Table 2, the relation between the molecular and motionless axis has minor effect on the cross section of this process.

Rotation of the molecular axis introduces a small contribution to the cross section of resonant charge exchange; the difference between cases a, b and d of Hund coupling, as well as for cases c and e is not essential for this process. According to Table 2, the difference of the cross sections for cases a and c of Hund coupling is not significant for atoms of groups III and VIII of the periodical table of elements, and it is essential for atoms of groups IV, V, VI, and VII. Transition between these coupling cases results from competition between the splitting $U(R)$ due to a long-range ion–atom interaction, the splitting $\Delta(R)$ due to the exchange interaction and the fine level splitting δ_f. Tables 5, 6 contain these values for atoms of groups III and VIII of the periodic table of elements.

The long-range splitting of levels depends on the atom and ion states. If atoms and ions are found in the ground states, the long-range splitting $U(R)$ of atomic levels for elements of groups III, IV, VI, VII of the periodic table results from interaction of the ion charge with the atom quadrupole moment and is given by

$$U(R) = \frac{5 \langle r^2 \rangle}{6R^3} \,, \tag{3.16a}$$

where R is the distance between interacting particles and $\langle r^2 \rangle$ is the mean square of the valent electron orbit in the atom. The long-range splitting of ion levels for elements of group IV, when the quadrupole momenta of the atom and ion is non-zero, is determined by interaction of quadrupole momenta, and the ion–atom long-range interaction potential $U(R)$ in this case is

$$U(R) = \frac{Q_i Q_a}{R^5} \,, \tag{3.16b}$$

where Q_i, Q_a are the quadrupole momenta of the atom and ion, respectively, which is $\pm 2 \langle r^2 \rangle / 5$ for states with zero orbital momentum projection and $\mp 4 \langle r^2 \rangle / 5$ for the states in which the orbital momentum projection onto the motionless axis is equal to 1. Expression (3.8b) relates to elements of groups IV and VII of the periodic table when the quadrupole moment of atoms and ions is non-zero. Next, the splitting of ion levels for elements of groups V and VIII, whose atoms have zero quadrupole moment, is given by

$$U(R) = \frac{12\alpha(\langle r^2 \rangle)^2}{25R^8} \qquad (3.16c)$$

where α is the atom polarizability. The value $\Delta\sigma/\bar{\sigma}$ in Tables 7 and 8 characterizes an error in the cross section due to the use of the exponential dependence for the exchange phase $\zeta(\rho) \sim \exp(-\gamma\rho)$ only, as has been done in Table 5.

Tables 7 and 8 demonstrate the role of different interactions for the resonant charge exchange process involving real ions and atoms. These Tables imply that a long-range splitting of molecular terms is important for elements of groups three and is negligible compared to the exchange interaction potential for molecular ions of rare gases. In addition, Table 8 contains the average cross sections of the resonant charge exchange processes for elements with valent p-electrons.

Let us consider the charge exchange process of rare gas atoms and ions if ions are initially found in the ground state ($j = 3/2$). Then at small collision velocities only the ground ion state participates in this process and the transition into the ion state $j = 1/2$ is forbidden. At high collision velocities this channel opens up and the resonant charge exchange process corresponds to the case a of Hund coupling. Let us assume that these coupling schemes lead to an identical cross section so that a variation of the cross section in the course of transition between the cases c and a of Hund coupling is due to different atom ionization potentials with the formation of different fine ion states. The jump in the cross section due to this effect is

$$\Delta\bar{\sigma}_{\text{res}} = \frac{1}{3}\frac{\Delta I}{I}\bar{\sigma}_{\text{res}} , \qquad (3.17)$$

where the first factor is the probability of the ion state $j = 1/2$, the second factor accounts for dependence (2.2) of the cross section on the electron binding energy, and ΔI is the difference in the ionization potentials for states with different total momenta that corresponds to the fine splitting of ion levels. According to this formula, the relative variation of the cross section is about 0.4% for Ar, about 2% for Kr, and about 4% for Xe. This effect was first observed experimentally in [40]. Collision velocity v for this transition can be estimated from

Table 7. Parameters of the resonant charge exchange process for collisions of atoms of group III of the periodic table with their ions at energy 1 eV in the laboratory frame.

	B	Al	Ga	In	Tl
$\bar{\sigma}$, 10^{-14} cm^2	1.1	1.8	2.0	2.2	2.0
γR_0	12	14	14	15	14
$\Delta\sigma/\bar{\sigma}$, %	0.7	0.5	0.4	0.3	0.4
δ_f, cm^{-1}	15	112	826	2213	7793
$U(R_0)$, cm^{-1}	360	350	320	330	390
$\Delta(R_0)$, cm^{-1}	11	5	3	2.5	2

Table 8. Parameters of the resonant charge exchange process for collisions of atoms of group VIII of the periodical table with their ions at energy 1 eV in the laboratory frame.

	Ne	Ar	Kr	Xe
$\bar{\sigma}$, 10^{-15} cm^2	3.3	5.8	7.5	9.8
γR_0	11	12	13	14
$\Delta\sigma/\bar{\sigma}$, %	0.8	0.5	0.4	0.3
δ_f, cm^{-1}	780	1432	5370	10537
$U(R_0)$, 10^{-3} cm^{-1}	5	4	2	2
$\Delta(R_0)$, cm^{-1}	13	8	5	3

the expression for the typical time of the process

$$\tau \sim \frac{1}{v}\sqrt{\frac{R_0}{\gamma}} \sim \Delta I. \tag{3.18}$$

As follows from this formula, the typical collisional energy for the transition between cases a and c of Hund coupling for the resonant charge exchange process is estimated to be ~ 10 eV for Ar, ~ 100 eV for Kr, and ~ 600 eV for Xe. At low energies the case c of Hund coupling is realized, i.e. the total electron momentum is the quantum number of this process.

In addition, the transitions between states with different total momenta are adiabatically forbidden at low energies and hence are practically absent, i.e. the ionic fine state does not change during collisions with atoms. The ratio of the cross sections $\sigma_{3/2}, \sigma_{1/2}$ of the resonant charge exchange for the total electron momenta $3/2$ and $1/2$ of ions is equal approximately to

$$\frac{\sigma_{1/2}}{\sigma_{3/2}} \approx \frac{I_{3/2}}{I_{1/2}} ,$$

where $I_{3/2}, I_{1/2}$ are the atom ionization potentials with the formation of the ion states with the total electron momenta $3/2$ and $1/2$ correspondingly. The relative difference between the cross sections of electron transfer for different fine states of ions is approximately 1%, 5% and 11% for argon, krypton and xenon, respectively.

3.4. Average cross sections of resonant charge exchange

Summarizing the above results, we conclude that the asymptotic theory is valid if colliding atoms and ions are found in the ground states or lower excited states. The asymptotic theory provides an accuracy of $1-5\%$ [28] for s-electron transitions at eV collision energies. In the case of p-electron transitions the asymptotic theory leads to an accuracy of better than 10% for the cross sections of resonant charge exchange averaged over the momentum directions. Table 9 lists the parameters for average cross sections of resonant charge exchange with p-electron transitions [29]. In this case the dependence of the cross section on the collision velocity is given by formula (2.2) which can be rewritten in the form

$$\frac{\sigma(v)}{\sigma(v_0)} = \left(\frac{v_0}{v}\right)^\delta , \quad \text{where } \delta = -\frac{d\ln\sigma}{d\ln v} = \frac{1}{2R_0\gamma} .$$

Table 9 gives the parameters of this formula at a collision energy of 1 eV in the laboratory frame.

Figures 4 and 5 present the average cross sections of the resonant charge exchange process in slow collisions for all elements of the

B. M. Smirnov

Table 9. Parameters of the average cross section of resonant charge exchange for elements with valent p-electrons of atoms and ions at a collision energy of 1 eV [29].

Element	B	C	N	O	F	Ne	Al
σ, 10^{-15} cm^2	11	8.6	6.2	6.6	4.9	3.3	19
$\alpha = -d\ln\sigma/d\ln v$	0.16	0.16	0.16	0.16	0.17	0.18	0.15
Element	Si	P	S	Cl	Ar	Ga	Ge
σ, 10^{-15} cm^2	15	11	10	8.0	5.8	20	18
$\alpha = -d\ln\sigma/d\ln v$	0.14	0.14	0.15	0.15	0.16	0.14	0.13
Element	As	Se	Br	Kr	In	Sn	Sb
σ, 10^{-15} cm^2	13	13	10	7.5	22	19	17
$\alpha = -d\ln\sigma/d\ln v$	0.14	0.14	0.13	0.15	0.14	0.13	0.13
Element	Te	I	Xe	Tl	Pb	Bi	
σ, 10^{-15} cm^2	16	13	10	21	20	22	
$\alpha = -d\ln\sigma/d\ln v$	0.13	0.13	0.14	0.14	0.13	0.12	

periodic table [28, 39]. In addition, as an illustration of the asymptotic theory, we plot in Fig. 6 experimental data for this process with the participation of krypton atom and ion. Note that the accuracy of the asymptotic theory for the cross sections of resonant charge exchange is better than measurement errors with the possible exception of data obtained from the mobility of ions in parent gases. Therefore we use experimental data for the analysis of the ion mobilities. The data of Figs 4 and 5 provide the continuation [49 – 51] for tables of the cross sections of this process for various elements. Note that in contrast to other models [52 – 54], where the transferring electron is modelled by an s-electron, we now account for coupling of momenta in this process within the framework of the asymptotic theory representing the cross section as an expansion in power series of a small parameter.

3.5. Resonant charge exchange at ultralow energies

In considering the resonant charge exchange process, we assumed linear trajectories of the colliding ion and atom. But at low collisional energies the distortion of a linear trajectory is of importance, and below we take this into account for the ion–atom polarization interaction potential $U(R) = -\alpha e^2/(2R^4)$ at large separations R (α is the atom polarizability, and e is the electron charge). In the low collisional velocity limit, the resonant charge exchange cross section is one half of the capture cross section as a result of ion–atom collision [55]

$$\sigma_{\text{res}}(v) = \frac{1}{2}\sigma_{\text{cap}}(v) = \pi\sqrt{\frac{\alpha e^2}{2\varepsilon}}, \qquad (3.19)$$

where $\varepsilon = \mu v^2/2$ is the energy of colliding particles in the center-of-mass reference frame, such that v is the collision velocity, μ is the ion–atom reduced mass. Indeed, the resonant charge exchange proceeds after ion–atom approach as a result of their capture and the probability of electron transfer is one half on average.

In the other limiting case we take into account a weak distortion of the linear trajectory. Since the electron transition mostly proceeds at closest approach, in the expression for the transition probability it is necessary to replace the impact collision parameter ρ along a linear trajectory by the distance of closest approach r_{min} for the distorted trajectory. The relation between these parameters for the polarization interaction potential has the form [55]

$$\rho^2 = r_{\text{min}}^2\left(1 + \frac{\alpha e^2}{2r_{\text{min}}^4\varepsilon}\right). \qquad (3.20)$$

From this we obtain for the resonant charge exchange cross section at weak distortion of the trajectory

$$\sigma_{\text{res}} = \frac{\pi}{2}R_0^2 + \frac{\pi}{4}\frac{\alpha e^2}{R_0^2\varepsilon}, \qquad (3.21)$$

where $\pi R_0^2/2$ is the resonant charge exchange cross section in the case of linear trajectories, and the parameter R_0 is given by formula (3.2) for non-degenerate ground states of colliding ion and atom.

PERIODIC SYSTEM OF ELEMENTS

Period \ Group	I	II	III	IV	V
1	1s $^2S_{1/2}$ *1.008* $_1$H 6.12 4.82 Hydrogen 3.65	1s^2 $^1S_0$ *4.003* 3.4 2.7 2.1 $_2$He Helium			
2	2s $^2S_{1/2}$ *6.491* $_3$Li 26 22 Lithium 18	2s^2 $^1S_0$ *9.012* $_4$Be 13 10 Berillium 8.3	2p $^2P_{1/2}$ *10.81* 14 12 9.0 $_5$B Boron	2p^2 $^3P_0$ *12.011* 9.9 8.2 6.6 $_6$C Carbon	2p^3 $^4S_{3/2}$ *14.007* 7.7 6.4 5.2 $_7$N Nitrogen
3	3s $^2S_{1/2}$ *22.990* $_{11}$Na 31 26 Sodium 21	3s^2 $^1S_0$ *24.305* $_{12}$Mg 18 15 Magnesium 12	3p $^2P_{1/2}$ *26.982* 14 12 9.0 $_{13}$Al Aluminium	3p^2 $^3P_0$ *28.086* 17 14 12 $_{14}$Si Silicon	3p^4 $^3P_2$ *30.974* 14 12 10 $_{15}$P Phosphorus
4	4s $^2S_{1/2}$ *39.098* $_{19}$K 40 34 Potassium 28	4s^2 $^1S_0$ *40.08* $_{20}$Ca 25 21 Calcium 17	3d4s^2 $^2D_{3/2}$ *44.956* $_{21}$Sc 24 20 Scandium 16	3d^{2}4s^2 $^3F_2$ *47.88* $_{22}$Ti 22 19 Titanium 15	3d^{3}4s^2 $^4F_{3/2}$ *50.942* $_{23}$V 23 19 Vanadium 16
	3d^{10}4s $^2S_{1/2}$ *63.546* 19 16 13 $_{29}$Cu Copper	4s^2 $^1S_0$ *65.38* 15 12 10 $_{30}$Zn Zinc	4p^2 $^2P_{1/2}$ *69.72* 26 22 18 $_{31}$Ga Gallium	4p^2 $^3P_0$ *72.59* 20 17 14 $_{32}$Ge Germanium	4p^3 $^4S_{3/2}$ *74.922* 15 13 11 $_{33}$As Vanadium
5	5s $^2S_{1/2}$ *85.468* $_{37}$Rb 45 38 Rubidium 32	5s^2 $^1S_0$ *87.62* $_{38}$Sr 29 25 Strontium 20	4d5s^2 $^2D_{3/2}$ *88.906* $_{39}$Y 25 21 Yttrium 18	4d^{2}5s^2 $^3F_2$ *91.22* $_{40}$Zr 23 20 Zirconium 16	4d^{4}5s $^6D_{1/2}$ *92.906* $_{41}$Nb 23 19 Niobium 16
	4d^{10}5s^2 $^2S_{1/2}$ *107.87* 20 17 14 $_{47}$Ag Silver	5s^2 $^1S_0$ *112.41* 16 14 11 $_{48}$Cd Cadmium	5p^2 $^2P_{1/2}$ *114.82* 28 24 20 $_{49}$In Indium	5p^2 $^3P_0$ *118.69* 22 19 15 $_{50}$Sn Tin	5p^3 $^4S_{3/2}$ *121.75* 20 17 14 $_{51}$Sb Antinomy
6	6s $^2S_{1/2}$ *132.90* $_{55}$Cs 51 44 Cesium 36	6s^2 $^1S_0$ *137.33* $_{56}$Ba 35 30 Barium 25	5d6s^2 $^2D_{3/2}$ *138.90* $_{57}$La 32 27 Lanthanum 23	5d^{2}6s^2 $^3F_2$ *178.49* $_{72}$Hf 22 18 Hafnium 15	5d^{3}6s^2 $^4F_{3/2}$ *180.95* $_{73}$Ta 20 17 Tantalium 14
	5d^{10}6s $^2S_{1/2}$ *196.97* 16 14 11 $_{79}$Au Gold	5d^{10}6s^2 $^1S_0$ *200.59* 14 12 10 $_{80}$Hg Mercury	6p^2 $^2P_{1/2}$ *204.38* 26 22 18 $_{81}$Tl Thallium	6p^2 $^3P_0$ *207.2* 22 19 16 $_{82}$Pb Lead	6p^3 $^4S_{3/2}$ *208.98* 26 22 19 $_{83}$Bi Bismuth
7	7s $^2S_{1/2}$ *[223]* $_{87}$Fr 53 45 Francium 38	7s^2 $^1S_0$ *226.02* $_{88}$Ra 35 30 Radium 25	6d7s^2 $^2D_{3/2}$ *227.03* $_{89}$Ac 33 28 Actinium 24		

Fig. 4. Resonant charge exchange cross sections for basic elements of the periodic table [28, 39].

Resonant charge exchange cross sections

Legend:
- Shell of valent electrons / Electron term
- Atomic weight
- Symbol
- Atomic number
- Element
- Example: 137.33 | 6s² ¹S₀ | 56 Ba | 35 / 30 / 25 | Barium
- Cross section of resonant charge exchange (in 10^{-15} cm² at 0.1, 1, 10 eV in laboratory frame)

Period	Group	Config / Term	Atomic weight	Cross sections (0.1, 1, 10 eV)	Z · Element
2	VI	$2p^4\,{}^3P_2$	15.999	8.3 / 6.9 / 5.6	8 O — Oxygen
2	VII	$2p^5\,{}^2P_{3/2}$	18.998	5.7 / 4.7 / 3.7	9 F — Fluorine
2	VIII	$2p^6\,{}^1S_0$	20.179	4.1 / 3.2 / 2.6	10 Ne — Neon
3	VI	$3p^4\,{}^3P_2$	32.06	13 / 10 / 8.6	16 S — Sulfur
3	VII	$3p^5\,{}^2P_{3/2}$	35.453	9.6 / 8.0 / 6.6	17 Cl — Chlorine
3	VIII	$2p^6\,{}^1S_0$	39.948	7.0 / 5.8 / 4.7	18 Ar — Argon
4	VI	$3d^5 4s\,{}^7S_3$	51.996	23 / 19 / 16	24 Cr — Chromium
4	VII	$3d^5 4s^2\,{}^6S_{5/2}$	54.938	20 / 17 / 14	25 Mn — Manganese
4	VIII	$3d^6 4s^2\,{}^5D_4$	55.847	19 / 16 / 13	26 Fe — Iron
4	VIII	$3d^7 4s^2\,{}^4F_{9/2}$	58.993	19 / 16 / 13	27 Co — Cobalt
4	VIII	$3d^8 4s^2\,{}^3F_4$	58.69	20 / 17 / 14	28 Ni — Nickel
4	VI	$4p^4\,{}^3P_2$	78.96	16 / 14 / 12	34 Se — Selenium
4	VII	$4p^5\,{}^2P_{3/2}$	79.94	12 / 10 / 8.2	35 Br — Bromine
4	VIII	$4p^6\,{}^1S_0$	83.80	9.0 / 7.5 / 6.2	36 Kr — Krypton
5	VI	$4d^5 5s\,{}^7S_3$	95.94	22 / 19 / 15	42 Mo — Molibdenum
5	VII	$4d^5 5s^2\,{}^6S_{5/2}$	[98]	21 / 18 / 14	43 Tc — Technetium
5	VIII	$4d^7 5s\,{}^5F_5$	101.07	21 / 18 / 14	44 Ru — Ruthenium
5	VIII	$4d^8 5s\,{}^4F_{9/2}$	102.91	20 / 17 / 14	45 Rh — Rhodium
5	VIII	$4d^{10}\,{}^1S_0$	106.42	18 / 16 / 13	46 Pd — Palladium
5	VI	$5p^4\,{}^3P_2$	127.60	18 / 16 / 13	52 Te — Tellurium
5	VII	$5p^5\,{}^2P_{3/2}$	126.90	14 / 12 / 10	53 I — Iodium
5	VIII	$5p^6\,{}^1S_0$	131.29	12 / 10 / 8.6	54 Xe — Xenon
6	VI	$5d^4 6s^2\,{}^5D_0$	183.85	20 / 17 / 14	74 W — Tungsten
6	VII	$5d^5 6s^2\,{}^6S_{5/2}$	186.21	20 / 17 / 14	75 Re — Rhenium
6	VIII	$5d^6 6s^2\,{}^5D_4$	190.2	18 / 15 / 13	76 Os — Osmium
6	VIII	$5d^7 6s^2\,{}^4F_{9/2}$	192.22	17 / 14 / 12	77 Ir — Iridium
6	VIII	$5d^9 6s^3\,{}^3D_3$	195.08	17 / 14 / 12	78 Pt — Platinum
6	VI	$6p^4\,{}^3P_2$	[209]	17 / 15 / 12	84 Po — Polonium
6	VII	$6p^5\,{}^2P_{3/2}$	[210]	18 / 15 / 13	85 At — Astatine
6	VIII	$6p^6\,{}^1S_0$	[222]	15 / 13 / 11	86 Rn — Radon

Actinides

Config / Term	Atomic weight	Cross sections (0.1, 1, 10 eV)	Z · Element
$6d^2 7s^2\,{}^3F_2$	232.04	28 / 24 / 20	90 Th — Thorium
$5f^2 6d7s^2\,{}^4K_{11/2}$	231.04	30 / 25 / 21	91 Pa — Protactinium
$5f^3 6d7s^2\,{}^6L_6$	238.03	28 / 24 / 20	92 U — Uranium
$5f^4 6d7s^2\,{}^6L_{11/2}$	237.05	46 / 39 / 33	93 Np — Neptunium
$5f^6 7s^2\,{}^7F_0$	[244]	38 / 32 / 27	94 Pu — Plutonium
$5f^7 7s^2\,{}^8S_{7/2}$	[243]	49 / 42 / 36	95 Am — Americium

Lantanides

Element	Shell	Atomic weight	Cross sections
58 Ce — Cerium	$4f5d6s^2\,{}^1G_4$	*140.12*	32 / 27 / 23
59 Pr — Praseodymium	$4f^3 6s^2\,{}^5I_4$	*140.91*	32 / 28 / 23
60 Nd — Neodymium	$4f^4 6s^2\,{}^5I_4$	*144.24*	32 / 27 / 23
61 Pm — Promethium	$4f^5 6s^2\,{}^6H_{5/2}$	*[145]*	32 / 27 / 22
62 Sm — Samarium	$4f^6 6s^2\,{}^7F_0$	*150.36*	31 / 26 / 22
63 Eu — Europium	$4f^7 6s^2\,{}^8F_{7/2}$	*151.96*	31 / 26 / 22
64 Gd — Gadolinium	$4f^7 5d6s^2\,{}^9D_2$	*157.25*	28 / 24 / 20
65 Tb — Terbium	$4f^9 6s^2\,{}^6H_{15/2}$	*158.92*	30 / 25 / 21
66 Dy — Dysprosium	$4f^{10} 6s^2\,{}^5I_8$	*162.50*	29 / 25 / 20
67 Ho — Holmium	$4f^{11} 6s^2\,{}^5I_{15/2}$	*164.93*	29 / 24 / 20
68 Er — Erbium	$4f^{12} 6s^2\,{}^3H_6$	*167.26*	28 / 24 / 20
69 Tm — Thulium	$4f^{13} 6s^2\,{}^2F_{7/2}$	*168.93*	28 / 23 / 19
70 Yb — Ytterbium	$4f^{14} 6s^2\,{}^1S_0$	*173.04*	27 / 23 / 19
71 Lu — Lutetium	$4f^{14} 5d6s^2\,{}^2D_{3/2}$	*174.97*	34 / 29 / 24

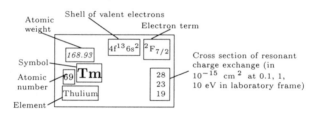

Atomic weight — Shell of valent electrons — Electron term
Symbol — Atomic number — Element

$4f^{13}6s^2$ — $^2F_{7/2}$ — *168.93* — 69 Tm — Thulium — 28 / 23 / 19

Cross section of resonant charge exchange (in 10^{-15} cm^2 at 0.1, 1, 10 eV in laboratory frame)

Fig. 5. Resonant charge exchange cross sections for lanthanides [28].

Above we assumed the electron transfer to proceed at the distance of closest approach and the resonant charge exchange cross section to be independent of the collision velocity. Along with changing the collision impact parameter by the distance of closest approach in the expression in the transition probability, it is necessary to replace the collision velocity v at large separations by the velocity $v_{\min}$ at the distance of closest approach [55]

$$v_{\min} = v\sqrt{1 + \frac{\alpha e^2}{2r_{\min}^4 \varepsilon}}\,.$$

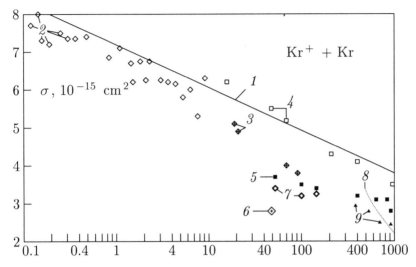

Fig. 6. The resonant charge exchange cross sections for krypton [39]. *1* — formulas (3.8a), (3.11), experiment: *2* — [41], *3* — [42], *4* — [43], *5* — [44], *6* — [45], *7* — [46], *8* — [47], *9* — [48].

This gives [56]

$$\sigma_{\text{res}}(v) = \frac{\pi}{2} R_0^2(v_{\min}) + \frac{\pi}{4} \frac{\alpha e^2}{R_0^2(v_{\min})\varepsilon}. \tag{3.22}$$

Since the cross section change due to distortion of collision trajectories is small in this limiting case and since the resonant charge exchange cross section is almost independent of the collision velocity, this velocity change is unimportant.

Sewing the limiting cases considered and accounting for weak velocity dependence of the cross section, we obtain the resonant charge exchange cross section in the form [56]

$$\sigma_{\text{res}} = \begin{cases} \dfrac{\pi}{2} R_0^2 + \dfrac{\pi}{4} \dfrac{\alpha e^2}{R_0^2 \varepsilon}, & \varepsilon \ge \dfrac{\alpha e^2}{2R_0^4}, \\[3mm] \pi \sqrt{\dfrac{\alpha e^2}{2\varepsilon}}, & \varepsilon \le \dfrac{\alpha e^2}{2R_0^4}. \end{cases} \tag{3.23}$$

Of course, the cross sections in the limiting cases coincide at the sewing point. The distortion of the collision trajectory for the resonant charge exchange process due to the polarization ion – atom interaction leads to an increase in the cross section. In particular, at the sewing point this increase is a factor of two. This effect also changes the cross section velocity dependence.

4.   Mobility of atomic ions in gases

4.1.   The character of ion drift in atomic gas

When an atomic ion moves in an atomic gas in an external electric field, it acquires an impetus from the electric field and returns it to gas atoms as a result of collisions. These collisions give rise to a friction force acting on the ion, so that the equation of motion for a test ion has the form

$$M \frac{d\mathbf{w}}{dt} = e\,\mathbf{F} - \frac{M\mathbf{w}}{\tau}. \tag{4.1}$$

Here M is the ion mass, e is the ion charge which is assumed to be equal to the electron charge, $\mathbf{w}$ is the average ion velocity, $\mathbf{F}$ is the electric field strength, and τ is the typical time of ion – atom collisions. The effect of collisions with atoms on the ion is then characterized by the parameter $\tau \sim 1/(N_{\mathrm{a}} v \sigma)$, where N_{a} is the number density of gas atoms, v is the typical ion – atom relative velocity, and σ is the typical cross section of ion – atom collisions responsible for change of the ion momentum in collisions.

Equation (4.1) can be a basis for the balance equation of ions in a gas subjected to an external electric field $\mathbf{F}$. If the number density of ions is small compared to the number density of atoms, and elastic collisions with atoms govern the character of the ion motion in the gas, in the stationary case the balance equation is

$$e\,\mathbf{F} = \frac{m\mathbf{w}}{\tau}, \tag{4.2}$$

under the conditions considered, where $\mathbf{w}$ is the average ion velocity and τ is a typical time between successive collisions of the ion with atoms. The left-hand side of this equation is the force on the ion from the electric field, and the right-hand side is the frictional force arising from collisions of the ion with the gas atoms. Below we shall derive the frictional force strictly, without using the tau approximation, but directly from the Boltzmann kinetic equation for ions.

The Boltzmann kinetic equation for ions moving in an atomic gas in an external electric field has the form

$$\frac{\partial f(\mathbf{v})}{\partial t} + e\,\mathbf{F}\,\frac{\partial f(\mathbf{v})}{\partial \mathbf{v}} = I_{\mathrm{col}}(f)\,, \tag{4.3}$$

where $f(\mathbf{v})$ is the velocity distribution function of ions, so that $\mathbf{v}$ is a current ion velocity, $I_{\mathrm{col}}(f)$ is the collision integral of a test ion with gas atoms. In the case of motion of an ion in an atomic gas we neglect inelastic ion–atom collisions. In addition, because of the small number density of ions we ignore ion–ion collisions and the presence of ions in the gas does not affect the equilibrium distribution function of ions. Therefore, the ion–atom collision integral takes the form [57]

$$I_{\mathrm{col}}(f) = \int \left[(f(\mathbf{v}')\varphi(v_{\mathrm{a}}') - f(\mathbf{v})\varphi(v_{\mathrm{a}})) \right] \mid \mathbf{v} - \mathbf{v}_{\mathrm{a}} \mid d\sigma dv_{\mathrm{a}}\,, \tag{4.4}$$

where $\mathbf{v}, \mathbf{v}'$ are the ion velocity before and after the collision, $\mathbf{v}_{\mathrm{a}}, \mathbf{v}_{\mathrm{a}}'$ are the atom velocities before and after the collision, $d\sigma$ is the collision differential cross section which leads to this change of velocities of the colliding particles, $\varphi(v_{\mathrm{a}})$ is the Maxwell distribution function of gas atoms.

Multiplying equation (4.3) with the collision integral (4.4) by $m\mathbf{v}$ and integrating over $d\mathbf{v}$ yields

$$e\,\mathbf{F}N_{\mathrm{i}} = \int M(\mathbf{v}' - \mathbf{v})g\,d\sigma f(\mathbf{v})\varphi(v_{\mathrm{a}})\,d\mathbf{v}d\mathbf{v}_{\mathrm{a}}\,. \tag{4.5}$$

Here N_{i} is the number density of ions and $\mathbf{g} = \mathbf{v} - \mathbf{v}_{\mathrm{a}}$ is the relative velocity of the colliding particles conserved in the collision. We employed the principle of detailed balance assuring the invariance of the system evolution under time reversal ($t \to -t$), yielding in this case

$$\int \mathbf{v} f(\mathbf{v})\varphi(v_{\mathrm{a}})g\,d\sigma d\mathbf{v}d\mathbf{v}_{\mathrm{a}} = \int \mathbf{v}' f(\mathbf{v})\varphi(v_{\mathrm{a}})g\sigma\,d\mathbf{v}d\mathbf{v}_{\mathrm{a}}\,.$$

Expressing the ion velocity $\mathbf{v}_1$ in formula (4.5) in terms of the relative ion–atom velocity $\mathbf{g}$ and introducing the center-of-mass velocity $\mathbf{V}$

by the relation

$$\mathbf{v} = \mathbf{g} + \frac{m_a}{M + m_a}\,\mathbf{V},$$

where m_a is the atom mass, we find $M(\mathbf{v}_1 - \mathbf{v}_1') = \mu(\mathbf{g} - \mathbf{g}')$, where

$$\mu = \frac{M m_a}{M + m_a}$$

is the reduced mass of colliding particles. The relative velocity after the collision can be written in the form $\mathbf{g}' = \mathbf{g}\cos\vartheta + \mathbf{k}\,g\sin\vartheta$, where ϑ is the scattering angle, and $\mathbf{k}$ is the unit vector perpendicular to $\mathbf{g}$. Because of the random distribution of $\mathbf{k}$ in the plane directed perpendicular to $\mathbf{g}$, the integration over scattering angles gives $\int(\mathbf{g} - \mathbf{g}')\,d\sigma = \mathbf{g}\,\sigma^*(g)$, where $\sigma^*(g) = \int(1 - \cos\vartheta)\,d\sigma$ is the diffusion cross section of ion-atom scattering. Thus the relation (4.5) takes the form [58, 59]

$$e\,\mathbf{F}N_i = \int \mu\,\mathbf{g}\,g\,\sigma^*(g)f(\mathbf{v})\varphi(v_a)\,d\mathbf{v}d\mathbf{v}_a \tag{4.6}$$

with the Maxwell distribution function of atoms on velocities

$$\varphi(v_a) = N_a \left(\frac{m_a}{2\pi T}\right)^{3/2} \exp\left(-\frac{m_a v_a^2}{2T}\right). \tag{4.7}$$

Relation (4.6) can serve as a basis for evaluation of the ion drift velocity. In the simplest Maxwell case $\sigma^*(g) \sim 1/g$, the right-hand side of this relation gives

$$\int \mathbf{g}f(\mathbf{v})\varphi(v_a)d\mathbf{v}d\mathbf{v}_a = (\mathbf{w} - \mathbf{w}_a)N_iN_a,$$

where $\mathbf{w}, \mathbf{w}_a$ are the average velocities of ions and atoms, correspondingly. Note that the distribution functions of ions $f(\mathbf{v})$ and atoms $\varphi(v_a)$ are normalized by the conditions

$$\int f(\mathbf{v})d\mathbf{v} = N_i, \quad \int \varphi(v_a)\,d\mathbf{v}_a = N_a$$

where N_i, N_a are the number densities of ions and atoms, respectively. In the case where the diffusion cross section $\sigma^*(g)$ is inversely proportional to the relative collision velocity g, the drift velocity of ions $\mathbf{w}$ in a motionless gas ($\mathbf{w}_a = 0$) yields

$$\mathbf{w} = \frac{e\mathbf{F}}{\mu\nu}, \quad \nu = N_a g\sigma^*(g). \tag{4.8}$$

The integral relation is useful for determination of the drift velocity of a heavy ion which is moving in a gas in an external electric field [60]. In this limiting case $M \gg m_a$ the ions have a narrow velocity distribution function compared to the atom distribution function, and we represent it in the form

$$f(\mathbf{v}) = N_i\delta(\mathbf{v} - \mathbf{w}), \quad \mathbf{w} = \frac{1}{N_i}\int \mathbf{v}f(\mathbf{v})\,d\mathbf{v}. \tag{4.9}$$

Hence, the relative ion–atom velocity is $\mathbf{g} = \mathbf{w} - \mathbf{v}_a$. Substituting this into relation (4.6) and integrating over angles, we obtain [60]

$$\frac{eF}{m_a N_a} = \frac{1}{w^2}\sqrt{\frac{2T}{m_a}}\exp\left(-\frac{m_a w^2}{2T}\right)\int_0^\infty \exp\left(-\frac{m_a g^2}{2T}\right)$$
$$\times g^2\sigma^*(g)dg\left(\frac{m_a wg}{T}\cosh\frac{m_a wg}{T} - \sinh\frac{m_a wg}{T}\right) \tag{4.10}$$

If the ion drift velocity w is small compared to the thermal atom velocity $\sqrt{T/m_a}$, this formula gives for the ion drift velocity

$$w = \left(\frac{2T}{m_a}\right)^{5/2}\frac{3\sqrt{\pi}eF}{8m_a N_a \int_0^\infty \exp\left(-\frac{m_a g^2}{2T}\right)g^5\sigma^*(g)dg}, \tag{4.11a}$$

$$w \ll \sqrt{\frac{2T}{m_a}}.$$

In the other limiting case we have from formula (4.10)

$$\frac{eF}{m_a N_a} = w^2\sigma^*(w), \quad w \gg \sqrt{\frac{2T}{m_a}}. \tag{4.11b}$$

Along with the integral relation for the average ion momentum (4.6) that follows from the kinetic equation for the distribution function of ions moving in a gas in an external electric field, a similar equation for the average ion energy is useful. In order to derive it, we multiply Boltzmann equation (4.3) by $Mv^2/2$ and integrate the result over ion velocities. Then the left-hand side of the resultant equation is the energy that is transmitted from the electric field to ions located in a unit volume and has the form $eFwN_i$. Repeating the operations we used in deriving equation (4.6), we obtain in this case

$$eFwN_i = \int \mu \mathbf{V}\mathbf{g}\, g\sigma^*(g) f(\mathbf{v}) \varphi(v_a)\, d\mathbf{v} d\mathbf{v}_a\,. \tag{4.12}$$

Below we demonstrate the use of this integral relation in the Maxwell case when the diffusion cross section $\sigma^*(g)$ is inversely proportional to the collision relative velocity g. Then using the relation

$$\mathbf{V}\mathbf{g} = \frac{M}{M+m_a}v^2 - \frac{m_a}{M+m_a}v_a^2 + \frac{M-m_a}{M+m_a}\mathbf{v}\mathbf{v}_a\,,$$

accounting for $\mathbf{w}_a = 0$ and expression (4.7) for the ion drift velocity, we obtain the ion mean energy in the form

$$\varepsilon_i = \frac{M}{2}\left\langle v^2 \right\rangle = \frac{(M+m_a)w_i^2}{2} + \frac{3}{2}T\,. \tag{4.13}$$

Above we have taken into account the expression for the mean kinetic energy of atoms

$$\frac{m_a}{2}\left\langle v_a^2 \right\rangle = \frac{3}{2}T\,.$$

Relation (4.13) may be used to evaluate the average energy of ions drifting in a gas in an external electric field. As an example, we consider the Wannier case [61] where the cross section of ion–atom scattering does not depend on the collision velocity and the scattering is isotropic in the center-of-mass reference frame. In this case the correct solution of this problem is possible at high fields yielding [61]

$$w_i = 1.147\sqrt{a\lambda}\,, \quad \varepsilon_i = 1.18Ma\lambda\,,$$

$$\text{where } a = \frac{eF}{M}\,, \quad \lambda = \frac{1}{N_a\sigma^*}\,, \quad a\lambda \gg T\,. \tag{4.14}$$

In this case $2\varepsilon_i/(Mw_i^2) = 0.90$ that characterizes the accuracy of formula (4.13) for this case.

4.2.   Mobility of ions at low field strengths

The mobility of a charged particle K is defined as the ratio of its drift velocity w to the electric field strength F, or

$$K = \frac{w}{F}. \tag{4.15}$$

The ion mobility does not depend on the electric field strength at low fields

$$eF\lambda \ll T, \tag{4.16}$$

where λ is the mean free path of ions in the gas. This condition implies that the energy which the ion takes from the electric field between subsequent collisions is small compared to its thermal energy. As a result, the ion drift velocity is small compared to the typical ion thermal velocity in this case. Equation (4.6) can be a basis for determination of the ion mobility in a gas at low fields. Then the distribution function is close to the Maxwell one and therefore can be written in the form

$$f(\mathbf{v}) = \varphi(v)\left[1 + v_x\psi(v)\right], \tag{4.17}$$

where $\varphi(v)$ is the Maxwell distribution function of the ion, the electric field is along the x-axis, and the function $\psi(v)$ can be determined by solving the kinetic equation.

A numerical method of solution of this equation is based on expansion of the function $\psi(v)$ in series of the Sonine polynomials $S_n^m(u^2)$, where $u = v/\sqrt{2T/M}$ is the reduced ion velocity. This method is called the Chapman–Enskog approximation [62, 63], and in the case of ion drift $n = 3/2$. We demonstrate the Chapman–Enskog method in the first approximation when ψ is independent of the ion velocity. Then the parameter ψ can be determined by integral relation (4.6) for the ion distribution function which in this case has the form

$$eFN_{\mathrm{i}} = h \int \mu g_x g\sigma^*(g) v_x \varphi(v)\varphi(v_{\mathrm{a}})d\mathbf{v}d\mathbf{v}_{\mathrm{a}}$$

$$= \frac{\psi}{3}\mu \int g^2\sigma^*(g)\varphi(v)\varphi(v_{\mathrm{a}})d\mathbf{v}d\mathbf{v}_{\mathrm{a}}.$$

Next, the ion drift velocity is by definition

$$w = \int \mathbf{v} f(\mathbf{v})\, d\mathbf{v} = \frac{\psi}{3} \left\langle v^2 \right\rangle .$$

From this we have for the ion mobility (4.15) in the first Chapman–Enskog approximation

$$K_1 = \frac{3e\sqrt{\pi}}{8 N_\mathrm{a} \bar\sigma \sqrt{2T\mu}} , \qquad (4.18\mathrm{a})$$

introducing in this way the average cross section as

$$\bar\sigma(T) = \frac{1}{2} \int\limits_0^\infty \sigma^*(x) e^{-x} x^2\, dx, \quad x = \frac{\mu g^2}{2T} . \qquad (4.18\mathrm{b})$$

Note that according to this definition, the average cross section $\bar\sigma(T)$ is the momentum transfer collision integral [62, 63] $\Omega^{(1,1)}(T)$: $\bar\sigma(T) \equiv \Omega^{(1,1)}(T)$. The first Chapman–Enskog approximation coincides with the correct result for the mobility of heavy ions in a gas $M \gg m_\mathrm{a}$ (compare formulas (4.11a) and (4.18)). In addition, this approach gives the correct result in the Maxwell case if $\sigma^*(g) \sim 1/g$ (compare formulas (4.8), (4.15) with (4.18)). The second Chapman–Enskog approximation yields [58]

$$K_{II} = K_I(1 + \Delta), \quad \Delta = \frac{2}{5} \frac{\left(m_\mathrm{a} \frac{d \ln \Omega}{dT}\right)^2}{m_\mathrm{a}^2 + 3M^2 + 2m_\mathrm{a} M \frac{\sigma^{(2)}}{\sigma^*}} , \qquad (4.19)$$

$$\Omega = \sqrt{\frac{2T}{m}}\, \bar\sigma , \quad \sigma^{(2)} = \int (1 - \cos^2 \vartheta)\, d\sigma .$$

The simplest case of ion drift in gases corresponds to the Maxwell case with the inverse velocity dependence for the diffusion cross section $\sigma^*(v) \sim 1/v$. Then the ion drift velocity is given by formula (4.7) that follows from integral relation (4.6) and is valid for any electric field strength. This case takes place for the polarization interaction potential between ion and atom

$$U(R) = -\frac{\alpha e^2}{R^4} , \qquad (4.20)$$

that is realized at large ion–atom distances R and α is the atom polarizability. The cross section σ_{cap} [55] of the ion capture by the atom for this interaction potential is given according to formula (3.19) by

$$\sigma_{cap} = 2\pi \sqrt{\frac{\alpha e^2}{\mu g^2}}, \qquad (4.21)$$

where μ is the ion–atom reduced mass, g is the relative ion–atom velocity. On the basis of this cross section we obtain from formulas (4.8), (4.11) for the mobility of ions in a foreign gas

$$K = \frac{e}{2\pi N_a \sqrt{\alpha \mu}}. \qquad (4.22)$$

The diffusion cross section of the ion–atom scattering under polarization interaction potential (4.20) between them is close to the capture cross section and is equal to [64, 2]

$$\sigma^*(v) = \int (1 - \cos \vartheta)\, d\sigma = 2.21\pi \sqrt{\frac{\alpha e^2}{\mu g^2}}, \qquad (4.23)$$

i.e. exceeds the capture cross section by about 10%. For this cross section the ion mobility reduced to the number density of gas atoms $N_a = 2.69 \cdot 10^{19}$ cm^{-3} is given by [64, 65]

$$K = \frac{36}{\sqrt{\alpha \mu}} \frac{\text{cm}^2}{\text{V} \cdot \text{s}}, \qquad (4.24)$$

where the ion–atom reduced mass is expressed in atomic mass units ($1.66 \cdot 10^{-24}$ g), and the polarizability is given in atomic units. It is of importance that this mobility does not depend on temperature, and ion parameters are taken into account by the ion–atom reduced mass only.

In reality, the accuracy of this formula may be estimated by matching this formula with experimental data, and Table 10 contains the ratio of the experimental quantities of the mobility of alkali ions in rare gases at room temperature [66, 67] to those calculated by formula (4.24). One can see that the experimental data exceed the theoretical ones, with the average ratio of these values being 1.15±0.10,

Table 10.   The ratio between experimental and
theoretical mobilities of ions in rare gases [2].

	He	Ne	Ar	Kr	Xe
Li	1.27	1.17	1.06	1.07	1.06
Na	1.38	1.24	1.07	1.06	1.09
K	1.35	1.28	1.08	1.10	1.07
Rb	1.29	1.26	1.08	1.08	1.07
Cs	1.29	1.18	1.07	1.08	1.07

and for the mobility of ions in helium this ratio is 1.30 ± 0.07. Correcting formula (4.24) on the basis of the data of Table 10, instead of formula (4.24) we approximately obtain

$$ K = \frac{31 \pm 2}{\sqrt{\alpha\mu}} \frac{\text{cm}^2}{\text{V} \cdot \text{s}}, \qquad (4.25) $$

Roughly, this corresponds to the use of polarization capture cross section (4.21) instead of diffusion cross section (4.23) for the polarization interaction potential in formula (4.8) for the ion drift velocity or its mobility. Thus, as follows from this analysis, the use of the polarization ion – atom interaction potential allows us to determine the ion mobility in gases at low fields with the accuracy about 10%.

4.3.   Mobility of ions in parent gases at low fields

We now consider the mobility of atomic ions in a parent atomic gas if the cross section of the resonant charge exchange process exceeds significantly the cross section of elastic ion – atom scattering. Then the effective ion scattering proceeds due to resonant charge exchange as a result of the Sena effect, as shown in Fig. 7. The scattering results in a transfer of charge from one atomic core to another, and hence the charge exchange cross section characterizes the ion mobility. As seen in the center-of-mass frame, this process leads to effective ion scattering by the angle $\vartheta = \pi$. Correspondingly, the diffusion cross section is [68]

$$\sigma^* = \int (1 - \cos\vartheta)\, d\sigma = 2\sigma_{\text{res}}, \qquad (4.26)$$

where σ_{res} is the cross section of the resonant charge exchange process. Then, assuming the resonant transfer cross section σ_{res} to be independent of the collision velocity, we obtain from formula (4.14) for the drift velocity of ions in the parent gas in the first Chapman–Enskog approach [60, 69, 2]

$$w_{\text{I}} = K_1 F = \frac{3\sqrt{\pi}eF}{16N\sigma_{\text{res}}\sqrt{TM}} = \sqrt{\frac{2T}{M}} \cdot 0.47\beta\,,$$

$$\beta = \frac{eF\lambda}{T} = \frac{eF}{2N_a\sigma_{\text{res}}T} \ll 1\,, \qquad (4.27)$$

where $\lambda = (2N_a\sigma_{\text{res}})^{-1}$ is the mean free path of ions with respect to the resonant charge exchange process, $M = m_a$, and β is the small parameter in this case. In the second Chapman–Enskog approximation we have in this case ($m_a = M$, $\sigma^{(2)} = 0$) according to formula (4.19) $\Delta = 1/40$. This gives for the ion velocity

$$w_{\text{II}} = \sqrt{\frac{2T}{M}} \cdot 0.48\beta\,, \quad \beta \ll 1. \qquad (4.28)$$

Above we assumed the resonant charge exchange cross section independent of the collision velocity. Following the method of papers [70, 60], we now account for weak velocity dependence (2.2) for the resonant charge exchange cross section. It is simpler to use velocity dependence (3.5) for the electron transfer cross section. Then on the basis of formulas (4.26) and (3.5), we have for average cross section (4.18b)

$$\bar{\sigma}(T) = \frac{1}{2} \int_0^\infty \sigma^*(x) e^{-x} x^2\, dx = \int_0^\infty \sigma_{\text{res}}(x_0) \left(\frac{x_0}{x}\right)^{\delta/2} e^{-x} x^2\, dx$$

$$= \sigma_{\text{res}}(x_0) x_0^{\delta/2} \Gamma\left(3 - \frac{\delta}{2}\right)\,.$$

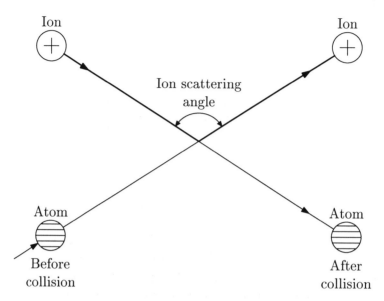

Fig. 7. The Sena effect.

By expanding this value in small δ and retaining two terms of this expansion, we represent the result in the form

$$\bar{\sigma}(T) = 2\sigma_{\text{res}}(x_0), \quad \text{where} \quad \ln x_0 = -\psi(3) = \frac{3}{2} - C = 0.923,$$

and $C = -0.577$ is Euler's constant. The resonant charge exchange cross section in formula (4.27) is taken at the collision energy $T\exp(0.923) = 2.5\,T$ in the center-of-mass reference frame, or at the double energy in the laboratory frame. The ion mobility in a parent gas in the limiting case of low electric field strengths according to the first Chapman–Enskog approximation [70] is

$$K_{\text{I}} = \frac{0.331e}{N_{\text{a}}\sqrt{Tm_{\text{a}}}\,\sigma_{\text{res}}(2.2v_T)} \qquad (4.29)$$

where the collision velocity is $v_T = \sqrt{2T/m_{\text{a}}}$, and the argument of the resonant charge exchange cross section σ_{res} indicates the velocity at which this cross section is taken. In the second Chapman–Enskog approximation we have [69, 2]

$$K_{\rm I} = \frac{0.341e}{N_{\rm a}\sqrt{Tm_{\rm a}}\,\sigma_{\rm res}(2.1\,v_T)}. \tag{4.30}$$

Reducing the ion mobility to the number density of gas atoms under normal conditions $N = 2.69 \cdot 10^{19}$ cm^{-3}, we rewrite formula (4.30) in the form [69, 2]

$$K = \frac{1340}{\sqrt{Tm_{\rm a}}\,\sigma_{\rm res}(2.1\,v_T)}\,\frac{{\rm cm}^2}{{\rm V}\cdot{\rm s}} \tag{4.31}$$

where the temperature T is given in Kelvins, the atom and ion mass $m_{\rm a} = M$ are expressed in atomic mass units, and the resonant charge exchange cross section is given in 10^{-15} cm^2; the argument indicates the collision velocity that corresponds to the collision energy $4.5T$ in the laboratory frame. Figure 8 shows the values of the mobilities of atomic ions in parent gases at temperatures $T = 300$ K, 800 K. These values are obtained on the basis of the resonant charge exchange cross section of Fig. 4 with accounting for correction due to ion–atom elastic scattering. The accuracy of these data is approximately 10%. Note that according to formula (4.31), the temperature dependence of the reduced mobility of ions in a parent gas is limited to neglecting the ion–atom elastic scattering

$$K(T) \sim T^{\delta/2-1/2}. \tag{4.32}$$

Above we neglected elastic scattering of the colliding ion and atom in the resonant charge exchange process. We now take account for this effect in the limit when elastic scattering gives a small contribution to the ion mobility. We have from formula (3.21)

$$\sigma_{\rm res} = \frac{\pi}{2}\,R_0^2 + \frac{\pi}{4}\,\frac{\alpha e^2}{R_0^2 \varepsilon},$$

where ε is the collision energy in the center-of-mass frame. Note that above we used only the first term of this formula, and the second term is considered here as a small correction. Averaging this cross section

	Shell of valence electrons		
n	ns	ns^2	np^6
1	*1.008* 12 8.4 $_1$H 0.31 Hydrogen 0.58	11 7.8 *4.003* 0.28 $_2$He 0.52 Helium	
2	*6.491* 1.1 0.74 $_3$Li 0.027 Lithium 0.051	*9.012* 1.8 1.3 $_4$Be 0.047 Berillium 0.091	4.0 2.8 *20.179* 0.10 $_{10}$Ne 0.19 Neon
3	*22.990* 0.92 0.63 $_{11}$Na 0.024 Sodium 0.044	*24.305* 0.80 0.56 $_{12}$Mg 0.021 Magnesium 0.038	1.6 1.1 *39.948* 0.041 $_{18}$Ar 0.077 Argon
4	*39.098* 0.29 0.20 $_{19}$K 0.0074 Potassium 0.014	*40.08* 0.44 0.31 $_{20}$Ca 0.011 Calcium 0.021	
4	0.49 0.33 *63.546* 0.013 $_{29}$Cu 0.023 Copper	0.58 0.42 *65.38* 0.015 $_{30}$Zn 0.029 Zinc	0.86 0.60 *83.80* 0.022 $_{36}$Kr 0.042 Krypton
5	*85.468* 0.18 0.12 $_{37}$Rb 0.0045 Rubidium 0.0082	*87.62* 0.27 0.18 $_{38}$Sr 0.0071 Strontium 0.013	
5	0.36 0.24 *107.87* 0.0090 $_{47}$Ag 0.016 Silver	0.41 0.28 *112.41* 0.011 $_{48}$Cd 0.020 Cadmium	0.52 0.36 *131.29* 0.013 $_{54}$Xe 0.025 Xenon
6	*132.90* 0.12 0.084 $_{55}$Cs 0.0032 Cesium 0.0058	*137.33* 0.17 0.12 $_{56}$Ba 0.0045 Barium 0.0083	
6	0.31 0.21 *196.97* 0.0079 $_{79}$Au 0.015 Gold	0.36 0.25 *200.59* 0.0094 $_{80}$Hg 0.017 Mercury	0.31 0.23 [222] 0.0080 $_{86}$Rn 0.016 Radon

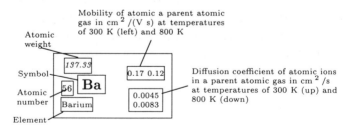

Fig. 8. The zero-field mobilities and diffusion coefficients of atomic ions in parent gases at temperature 300 K and 800 K.

over collision velocities, the mean cross section or the momentum transfer collision integral in the expression of the ion mobility reads:

$$\bar{\sigma}(T) \equiv \Omega^{(1,1)}(T) = \frac{2}{1+\Delta} \cdot \left[\sigma_{\text{res}}(4.5\,T) + \frac{\pi^2 \alpha e^2}{16 T \sigma_{\text{res}}(3\,T)} \right], \quad (4.33)$$

and the argument indicates the energy in the laboratory frame at which the resonant charge exchange cross section is evaluated (in this reference frame the colliding atom is at rest). The correction of the second Chapman – Enskog approximation according to formula (4.19) and velocity dependence (3.5) for the charge exchange cross section yield:

$$\Delta = \frac{(1-\delta)^2}{40}.$$

Thus from formula (4.33), if $\Delta K \ll K$ (the total mobility is $K_0 - \Delta K$, and K_0 is the mobility in neglecting ion – atom elastic scattering), we obtain for the mobility correction

$$\frac{\Delta K}{K} = \frac{\pi^2 \alpha e^2}{16\, T \sigma_{\text{res}}(3\,T) \sigma_{\text{res}}(4.5\,T)} . \quad (4.34)$$

The value of this effect can be understood from Table 11 [2] which presents a relative decrease ΔK of the ion mobility in the parent gas due to elastic ion – atom scattering determined by the polarization interaction. This correction is given in Table 9 for ions and atoms of the alkali metals and rare gases. One can see that the contribution of ion – atom elastic scattering into the mobility exceeds the accuracy of this value determination using the asymptotic theory for the resonant charge exchange cross section. Hence, we account for this effect, including the data of Fig. 8. Note that the temperature dependence of the correction to the ion mobility and of its relative value are

$$\Delta K(T) \sim T^{3\delta/2-3/2} , \quad \frac{\Delta K}{K} \sim T^{\delta-1} , \quad (4.35)$$

respectively. The effect of elastic ion – atom scattering is taken into account in Fig. 8, as well as in Fig. 9, where the mobility of atomic helium ions in helium at low electric field strengths evaluated using formula (4.24) is compared with experimental data [71 – 75].

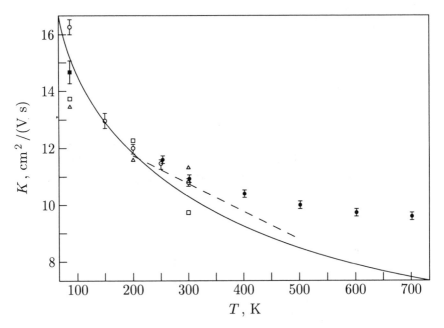

Fig. 9. The mobility of He$^+$ in helium at low fields. Theory: *1* — formulas (4.23), (4.24), *2* — [64] experiment: 3 — [71], 4 — [72], 5 — [73], 6 — [74], 7 — [75].

Table 11. The relative decrease of the ion mobility $\Delta K/K$ in the parent gases at room temperature due to elastic ion – atom scattering [2].

Ion, gas	$\Delta K/K$, %	Ion, gas	$\Delta K/K$, %
He	5.8	Li	12
Ne	7.8	Na	8.2
Ar	11	K	8.9
Kr	6.0	Rb	7.7
Xe	9.2	Cs	7.5

Table 12. The zero-field mobility $(\text{cm}^2/\text{V s})$ of inert gas ions in the parent gases at indicated temperatures. Theoretical data are given in parentheses, experimental data are averaged over different measurements.

T, K	≈ 300	195	77
He	10.4 ± 1.0 (11)	12 ± 1(13)	16 ± 1(17)
Ne	4.1 ± 0.1 (4.0)	4.3 ± 0.3 (4.6)	5.3 ± 0.5 (6.0)
Ar	1.5 ± 0.1 (1.6)	1.9 ± 0.3 (1.8)	2.1 ± 0.1 (2.3)
Kr	0.85 ± 0.1(0.81)	(0.92)	(1.1)
Xe	0.53 ± 0.1 (0.52)	(0.60)	(0.78)

Evaluating the correction to the mobility due to elastic scattering, we are able to calculate the mobility in a wide temperature range compared to room temperature. Table 12 contains averaged experimental data for the zero-field mobilities of the inert gas ions in parent gases taken from [5]. We compare them with calculations based on formulas (4.31), (4.33) and the asymptotic theory for the resonant charge exchange cross section. Note that the accuracy of the theory and experiment is comparable for these systems and we estimate it to be better than 10%. The comparison given in Table 10 confirms this accuracy. We provide one more example of comparison of the theoretical and experimental data. Averaged over measurements [76 – 79], the zero-field mobility of atomic cesium ions in cesium vapors is $0.12 \pm 0.03 \text{ cm}^2/\text{V s}$ in the temperature range 560 ± 80 K, and the theory gives the mobility $0.11 \pm 0.01 \text{ cm}^2/\text{V s}$ for this case and the temperature range. Hence the theory provides a higher accuracy than the experiment.

A more precise comparison of the asymptotic theory and experiment is given in Fig. 11 and 12, where the values of the momentum transfer collision integral evaluated by formula (4.33) are compared with those obtained by Viehland and Mason [8] from the experiments. In this case the momentum transfer collision integral is determined from the mobilities calculated by formula (4.18a), and the special analysis allows one to determine the effective temperature for the cross section. This comparison in the helium case (Fig. 10) shows that formula (4.33) is valid down to small temperatures where the contribution from elastic scattering to the momentum transfer collision integral is not small. Next, the resonant charge exchange cross sections for the xenon ions (Fig. 11) which are found in different fine states differ mostly due to different binding energies of the transferring electron, and the difference in these cross sections is approximately 10%. From the above comparison it follows that the asymptotic theory for the resonant charge exchange process provides the accuracy of the calculated mobilities of atomic ions in parent gases within 10%.

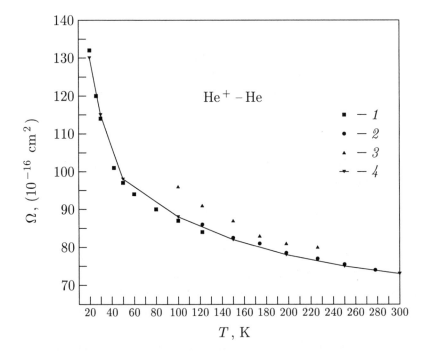

Fig. 10. The momentum transfer collision integral, obtained in [8] with usage of mobility experimental data: *1* — [80, 81], *2* — [82], *3* — [80], *4* — formula (4.33).

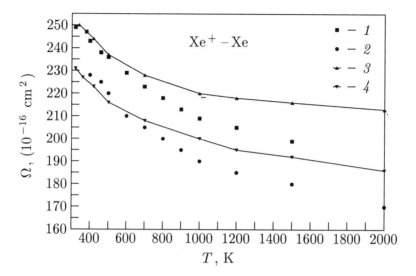

Fig. 11. The momentum transfer collision integral for atomic xenon ions in different fine states. *1,2* are obtained for the total ion momentum 3/2 and 1/2, correspondingly in [8] on the basis of measurements [83]; *3,4* are given by formula (4.33) for the ion momentum 3/2 and 1/2, respectively.

4.4. Mobility of ions at intermediate and high field strengths

At high electric fields, the condition (4.16) is reversed and takes the form

$$eF\lambda \gg T. \tag{4.36}$$

In that case the ion drift velocity is much greater than its thermal velocity, and because of the absence of elastic scattering, ion velocities along the field direction are much larger than in other directions. We start from ion drift in a foreign gas when a long-range interaction between atoms is the polarization one. The typical ion–atom interaction potential is given in Fig. 12, and a long-range tail of this interaction is the polarization interaction potential (4.20). A typical ion energy is small compared to the potential well depth and the ion drift velocity is determined by the polarization interaction potential. According to formula (4.8), the ion mobility does not depend on the

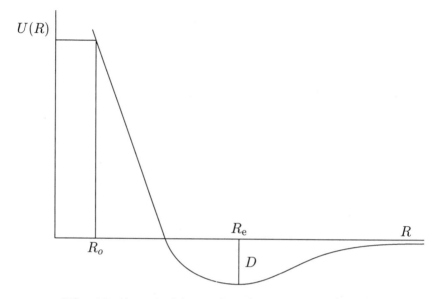

Fig. 12. A typical form of the ion-atom interaction
potential as a function of the separation.

electric field strength. On contrary, if we have the dependence of the
ion mobility on the mean ion energy or the electric field strength, it
is constant at small ion energies or fields, and the mobility obtains
the dependence on these quantities when a typical ion energy is comparable with the potential well depth. An example of such a mobility
dependence is given in Fig. 13. In principle, this dependence gives
the interaction potential as a function of the ion – atom distance. For
example, we approximate the ion – atom interaction potential $U(R)$
as a function of ion – atom distances R by the formula

$$U(R) = \frac{D}{2} \left[(1 + \gamma) \left(\frac{R_e}{R} \right)^{12} - 4\gamma \left(\frac{R_e}{R} \right)^6 - 3(1 - \gamma) \left(\frac{R_e}{R} \right)^4 \right] ,$$

$$(4.37)$$

where D is the potential well depth, R_e is the equilibrium ion –
atom distance corresponding to the interaction potential minimum.
Table 13 contains the parameters of this interaction potential [84]
obtained from the mobility data. Note that the value $3(1 - \gamma)DR_e^4$
in this expression is the atom polarizability, and this parameter com-

Table 13.

Ion – atom	D, meV	R_e, Å	γ
$\mathrm{Li^+ - He}$	47.4	2.22	0.1
$\mathrm{K^+ - Ar}$	121	3.00	0.2
$\mathrm{Rb^+ - Kr}$	119	3.34	0.2
$\mathrm{Cs^+ - Xe}$	106	3.88	0.2

bination practically coincides with the atom polarizability. The data of Fig. 13 and Table 13 demonstrate the connection between the ion – atom interaction potential and the mobility of ions in an atomic gas. One can determine the ion – atom interaction potential on the basis of the mobility dependence on the electric field strength. This interaction potential can be used for calculation of the diffusion coefficients of ions moving in the gas in an external electric field, as well as the information on the ion diffusion coefficients may be used for determination of the ion – atom interaction potential, if the transport coefficients are determined by ion – atom elastic scattering (for example, [85, 86]).

Let us consider the drift of atomic ions in a parent gas at high fields. We neglect ion – atom elastic scattering in the course of ion evolution, and ion scattering is determined by the resonant charge exchange process. The sequence of events in a strong electric field is such that the ion accelerates under the action of the field, stops as a result of the charge exchange event, and then this process repeats. In considering this limiting case, and following to [11, 12], we find the ion drift velocity under the assumption that the charge exchange cross section σ_{res} does not depend on the collision velocity. Ions are accelerated in the electric field direction, and the velocity in the field direction is large compared to that in transverse directions. We introduce the distribution function $f(v_x)$ of ions on velocities v_x in the field direction which is analogous to the probability $P(t)$ of absence of the charge exchange event through time t after the

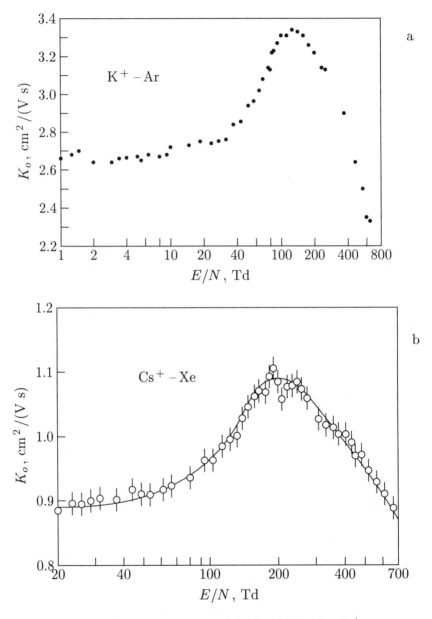

Fig. 13. The experimental drift velocity for K^+ –
Ar [87] (a) and Cs^+ – Xe [88] (b) as a function of
the reduced electric field strength.

previous exchange event. This probability satisfies the equation

$$\frac{dP}{dt} = -\nu P \,,$$

where $\nu = N v_x \sigma_{res}$ is the rate of the resonant charge exchange process. Its solution is

$$P(t) = \exp\left(-\int_0^t \nu \, dt'\right) \,.$$

The equation of motion for the ion,

$$M \frac{dv_x}{dt} = eF \,, \tag{4.38}$$

relates the ion velocity to the time after the last collision by the expression $v_x = eFt/M$, so that $P(t)$ is the velocity distribution function for the ions. Assuming the cross section of the resonant charge transfer process σ_{res} to be independent of the collision velocity, the distribution function is

$$f(v_x) = C \exp\left(-\frac{M v_x^2}{2eF\lambda}\right) \,, \quad v_x > 0 \,, \tag{4.39}$$

where C is a normalization factor, and the mean free path is $\lambda = 1/(N\sigma_{res})$. The ion drift velocity and the mean ion energy are

$$w = \overline{v_x} = \sqrt{\frac{2eF\lambda}{\pi M}} \,, \quad \bar{\varepsilon} = \frac{M \overline{v_x^2}}{2} = \frac{eF\lambda}{2} \tag{4.40}$$

respectively, and according to criterion (4.36), this drift velocity exceeds significantly the thermal velocity of atoms. Note that in this case we have

$$\frac{2\bar{\varepsilon}}{(M + m_a)w^2} = \frac{\pi}{4} \,,$$

whereas according to approximate relation (4.13) for this case this ratio is equal to one. This characterizes the accuracy of formula (4.13) in a general case.

The above operation allows us to take into account the energy dependence of the resonant charge exchange cross section. Indeed, let us use formula (3.5) for the resonant charge exchange cross section

$$\sigma_{\text{res}}(v) = \sigma_{\text{res}}(v_0) \left(\frac{v_0}{v}\right)^\delta , \quad \delta = \frac{2}{\gamma R_0}$$

where the argument gives the collision velocity, and we have $\delta \ll 1$. Repeating the above operations, we obtain for the distribution function in this case instead of formula (4.39)

$$f(v_x) = C \exp\left[-\frac{M v_x^2 N_a \sigma_{\text{res}}(v_x)}{(2-\delta)eF}\right] , \quad v_x > 0 , \qquad (4.41)$$

From this we obtain for the ion drift velocity w and the average kinetic ion energy $\bar{\varepsilon}$ instead of formula (4.40)

$$w = \left[\frac{(2-\delta)eF}{M N_a \sigma_{\text{res}}(v_0) v_0^\delta}\right]^{\frac{1}{2-\delta}} \frac{\Gamma\left(\frac{2}{2-\delta}\right)}{\Gamma\left(\frac{1}{2-\delta}\right)} ,$$

$$\bar{\varepsilon} = \frac{M}{2} \left[\frac{(2-\delta)eF}{M N_a \sigma_{\text{res}}(v_0) v_0^\delta}\right]^{\frac{2}{2-\delta}} \frac{\Gamma\left(\frac{3}{2-\delta}\right)}{\Gamma\left(\frac{1}{2-\delta}\right)} . \qquad (4.42)$$

In order to account for weak velocity dependence of the resonant charge exchange cross section, we use the same method as when deriving formula (4.29). Indeed, let us expand formulas (4.42) for small δ and require the term proportional δ be zero. This allows us to choose an appropriate value of v_0. Applying this operation for the drift velocity, from the above requirement we obtain the following equation

$$\ln\left[\frac{2eF}{M N_a \sigma_{\text{res}}(v_0) v_0^2}\right] = 1 - 2\psi(1) + \psi\left(\frac{1}{2}\right)$$

which gives $v_0 = 1.3\sqrt{\frac{eF\lambda}{M}}$.

Repeating these operations for the ion mean energy, we find the optimal velocity for the mean energy $v_0 = 1.4\sqrt{\frac{eF\lambda}{M}}$.

Thus, formula (4.42) can be rewritten in the form in the high field strength limit

$$w = \sqrt{\frac{2eF}{\pi M N_a \sigma_{\text{res}} \left(1.3\sqrt{\frac{eF\lambda}{M}}\right)}}, \quad \bar{\varepsilon} = \frac{eF}{2N_a \sigma_{\text{res}} \left(1.4\sqrt{\frac{eF\lambda}{M}}\right)}.$$

$$(4.43)$$

The drift ion velocity at moderate electric field strengths can be determined by solving the kinetic equation for the ion distribution function [89, 69, 90]. It is convenient to approximate the distribution function in the form

$$w = \sqrt{\frac{2T}{M}} \cdot \frac{0.48\beta}{(1 + 0.22\beta^{3/2})^{1/3}}, \quad (4.44a)$$

where β is the solution of the equation

$$\beta = \frac{eF}{2 T N_a \sigma_{\text{res}} \left[\sqrt{\frac{2T}{M}}(4.5 + 1.6\beta)\right]}. \quad (4.44b)$$

Because of weak velocity dependence of the resonant charge exchange cross section σ_{res}, this relation corresponds to the definition of the parameter β in formula (4.27). Note that the expansion of the ion mobility at low strengths has the form $K = K_0 + CF^2$, where C is a constant, and this expansion follows from the symmetry consideration. Formula (4.35a) gives the second term of expansion over a small parameter $\sim F^{3/2}$, and this is not correct. Because this is an approximate formula which does not pretend on the accuracy better than 10%, this inaccuracy is not significant or realistic.

Figure 14 compares calculations on the basis of formulas (4.35) with experimental data [91 – 93]. In addition, Fig. 15 shows the mobility of atomic argon ions in argon at small and moderate fields. This comparison with the experiment characterizes the accuracy of the asymptotic theory for the mobility of atomic ions in the parent gas or vapors. Thus, the resonant charge exchange process whose nature is connected with the structure of atomic particles is of importance for ion transport processes in parent gases. The asymptotic theory that accounts for the process nature, simplifies the analysis and allows one to evaluate the parameters of this process with a suitable accuracy under various real conditions.

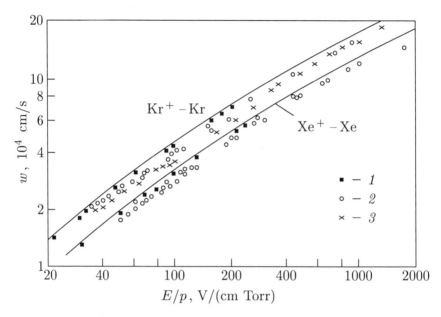

Fig. 14. The drift velocity of Kr^+ and Xe^+ in parent gases as a function of the reduced electric field strength. Solid curve — formula (4.44) [90], experiment: *1* — [91], *2* — [92], *3* — [93].

4.5. Diffusion of atomic ions in gases in external fields

Under thermodynamic equilibrium, the diffusion coefficient of ions is connected with their mobility by the Einstein relation [96 – 98]

$$D = \frac{KT}{e}. \tag{4.45}$$

In particular, formulas (4.18), (4.19) and the Einstein relation give the diffusion coefficient of ions in an atomic gas in the second Chapman – Enskog approximation [58]:

$$D_{II} = \frac{1}{(1 + \Delta)} \frac{3\sqrt{\pi T}}{8N\bar{\sigma}\sqrt{2\mu}}, \tag{4.46}$$

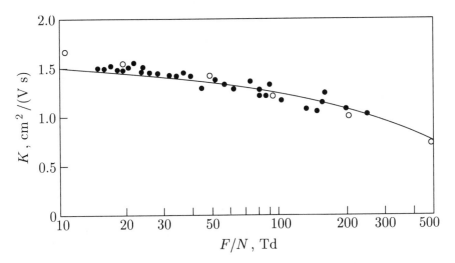

Fig. 15. The mobility of Ar^+- ions in argon. Theory: formula (4.31) — open circles. Experiment: closed circles — [94], solid curve — [95, 1].

where the average cross section of ion–atom scattering is given by formula (4.18b). The diffusion coefficient does not depend on the direction as far as the distribution function of ions is isotropic. But in strong electric fields the ion distribution function is anisotropic, and the diffusion coefficients for motion along the field $D_{\parallel}$ and perpendicular to it $D_{\perp}$ become different.

The diffusion coefficient of ions follows from formulas (4.25) and (4.31) for the ion mobilities at low fields and the Einstein relation. Using formula (4.25) for the diffusion coefficient of ions in a foreign gas, we get

$$D = (27 \pm 2) \cdot 10^{-4} \frac{cm^2}{V \cdot s} \cdot \frac{T}{\sqrt{\alpha\mu}},$$

and using formula (4.31) for the mobility of atomic ions in the parent gas, we have

$$D = 0.12 \frac{cm^2}{V \cdot s} \cdot \sqrt{\frac{T}{M}} \frac{1}{\sigma_{res}(2.1 v_T)} .$$

These diffusion coefficients relate to the normal number density of gas atoms $N = 2.69 \cdot 10^{19}$ cm^{-3}, the temperature T is given in Kelvins, and other notations are the same as in formulas (4.25) and (4.31).

In order to demonstrate this, we find the diffusion coefficients in the case of motion of atomic ions in a parent atomic gas in a strong electric field. The kinetic equation for the ion velocity distribution function f in a general case of ion motion in a gas in an external electric field has the following form

$$(\mathbf{v} - \mathbf{w})\nabla f + \frac{e\mathbf{F}}{M}\frac{\partial f}{\partial \mathbf{v}} = I_{\text{col}}(f) \tag{4.47}$$

and the ion diffusion coefficient in this gas D is defined as $\mathbf{j}_i = -D\nabla N_i$. Here $\mathbf{j}_i$ is the ion current density due to the ion number density gradient which is small on a distance of the order of the mean free path of ions in a gas. Hence, we expand the ion distribution function $f(\mathbf{v})$ over this small parameter by representing it in the form

$$f(\mathbf{v}) = f_0(\mathbf{v}) - \Phi(\mathbf{v})\nabla \ln N_i \,. \tag{4.48}$$

The unperturbed distribution function $f_0(\mathbf{v})$ satisfies the equation

$$\frac{e\mathbf{F}}{M}\frac{\partial f_0}{\partial \mathbf{v}} = I_{\text{col}}(f_0) \,.$$

In addition, the normalization condition gives

$$\int \Phi(\mathbf{v})\, d\mathbf{v} = 0 \,.$$

Substituting the above expansion for the ion distribution function into kinetic equation (4.47), we obtain the equation for the addition part of the distribution function

$$(v_z - w_z)f_0 = \frac{e\mathbf{F}}{M}\frac{d\Phi}{dv_z} + I_{\text{col}}(\Phi) \tag{4.49}$$

where z is the direction of the density gradient.

Let us determine the ion current density $\mathbf{j} = \int \mathbf{v} f(\mathbf{v}) d\mathbf{v}$ on the basis of the distribution function. Using the definition of the

diffusion coefficient, we then obtain the diffusion coefficient of ions in
the direction z

$$D_z = \frac{1}{N_i} \int (v_z - w_z)\Phi(\mathbf{v})\,d\mathbf{v}\,.\tag{4.50}$$

We now obtain an integral relation for the distribution func-
tion addition $\Phi(\mathbf{v})$ analogous to relation (4.6) by multiplying equation (4.50) by v_z and integrating it over ion velocities. Because of
the normalization condition for the addition $\Phi(\mathbf{v})$, the second term in
the right-hand side of the integral equation is zero, and this equation
takes the form

$$\left(\overline{v_z^2} - w_z^2\right) N_i = \frac{m_a}{M + m_a} \int g_z g\sigma^*(g)\Phi(\mathbf{v})\varphi(v_a)\,d\mathbf{v}\,d\mathbf{v}_a\tag{4.51}$$

where we used the operation for derivation of integral relation (4.6)
including the integration over scattering angles

$$\int (v_z - v_z')\,d\sigma = \frac{m_a}{M + m_a} g_z \sigma^*(g)$$

and we used the notations of formula (4.6). Equation (4.51) allows
us to find the diffusion coefficient in the case $\sigma^*(g) \sim 1/g$. Then this
equation contains the same integral from the addition distribution
function as expression (4.50) for the diffusion coefficient, and we have

$$D_z = \frac{M}{\mu\nu} \left(\overline{v_z^2} - w_z^2\right)\tag{4.52}$$

where the collision rate is $\nu = N_a g\sigma^*(g)$. One can see that for
the isotropic distribution function the diffusion coefficients of ions
in different directions are identical, while they are different for an
anisotropic distribution function in the ion rest frame.

We now determine the diffusion coefficient of ions along electric
field when atomic ions move in the parent gas in a strong electric
field. Assuming the mean free path of ions λ to be independent of
the ion velocity, we write the kinetic equation for ion distribution
function (4.47) in the form

$$\frac{eF}{M}\frac{\partial f}{\partial v_x} = \frac{1}{\lambda}\delta(v_x)\int v_x' f(v_x')\,dv_x' - \frac{v_x}{\lambda} f(v_x)$$

and introducing the reduced velocity

$$u = \left(\frac{M}{eF\lambda}\right)^{1/2} v_x,$$

we obtain equation (4.49) in the following form

$$\frac{2\lambda}{\pi w} N_{\mathrm{i}} \left(u - \sqrt{\frac{2}{\pi}}\right) \exp\left(-\frac{u^2}{2}\right) \eta(u) = \frac{d\Phi}{du} + u\Phi - \delta(u) \int u'\Phi(u')\, du',$$

where we use for the ion drift velocity $w = (2eF\lambda/\pi M)^{1/2}$. Solving this equation, we obtain

$$\Phi = \frac{2\lambda}{\pi w} N_{\mathrm{i}} \exp\left(-\frac{u^2}{2}\right) \left(\frac{u^2}{2} - \sqrt{\frac{2}{\pi}} + C\right) \eta(u)$$

and the constant C follows from the normalization condition $\int \Phi(u)du = 0$ that gives $C = 2/\pi - 1/2$. Thus, we finally obtain

$$D_\| = \frac{1}{N_{\mathrm{i}}} \int (v_x - w)\, \Phi\, dv_x = \lambda w \left(\frac{2}{\pi} - \frac{1}{2}\right) = 0.137\,\lambda w. \qquad (4.53)$$

The transversal diffusion coefficient of atomic ions in a parent gas can be readily found because the distribution function in the perpendicular direction to the field coincides with the Maxwell distribution function of atoms. The variables in the kinetic equation are divided for motion along the field and perpendicular to it, and the diffusion coefficient in the transversal direction to the field is equal to [59]

$$D_\perp = \frac{T\lambda}{Mw}. \qquad (4.54)$$

The ratio of the longitudinal and traversal diffusion coefficients in this case is:

$$\frac{D_\|}{D_\perp} = 0.137\frac{Mw^2}{T} = 0.087\frac{eF\lambda}{T} \qquad (4.55)$$

which is large because of criterion (4.36).

Let us consider the Lorentz case when the ion mass M is small in comparison with the atom mass m_{a}, $M \ll m_{\mathrm{a}}$ and the electric field strength is high. In particular, this case relates to motion of electrons in an atomic gas. If the rate of ion – atomic collisions $\nu(v)$ does not depend on the collision velocity v, then according to formula (4.52) we have

$$D_\perp = D_\| = \frac{\langle v^2 \rangle}{3\nu}, \quad w^2 \ll \langle v^2 \rangle, \quad \nu(v) = \text{const}. \tag{4.56}$$

If the cross section of ion – atom collision $\sigma^*(v)$ does not depend on the collision velocity v, the ion drift velocity w and the diffusion coefficients in the limit of high fields ($M \ll m_a$, $eF\lambda \gg T$) [61, 1, 99] are, respectively:

$$w = 0.897 \frac{\sqrt{eF\lambda}}{(m_a M)^{1/4}}, \quad D_\perp = 0.292 \left(\frac{m_a}{M} \right)^{1/4} \sqrt{\frac{eF}{M}} \lambda^{3/2},$$

$$D_\| = 0.144 \left(\frac{m_a}{M} \right)^{1/4} \sqrt{\frac{eF}{M}} \lambda^{3/2}. \tag{4.57}$$

We note that the results are expressed through the diffusion cross section and are identical for the classical and quantum character of ion – atom scattering. Hence, these formulas are valid both for electrons and ions.

We also consider the Wannier case [61] at high electric field strengths $eF\lambda \gg T$ when the cross section of ion – atom collision $\sigma^*(v)$ does not depend on the collision velocity v and the ion and atom masses are identical. In this case the ion drift velocity w, the transversal and longitudinal diffusion coefficients are, respectively:

$$w = 1.147 \sqrt{\frac{eF\lambda}{M}}, \quad D_\perp = 0.320 \sqrt{\frac{eF}{M}} \lambda^{3/2}, \quad D_\| = 0.220 \sqrt{\frac{eF}{M}} \lambda^{3/2}. \tag{4.58}$$

It is convenient to use the modified model of hard spheres [100] in the case of the classical ion – atom scattering with the interaction potential $U(R) = \text{const}/R^n$, $n \gg 1$. This interaction takes place when an ion penetrates into the atomic core, and transport coefficients are the expansions over a small parameter $1/n$. Then we represent the transport coefficients of ions at high electric field strengths in the form [5]

$$w = A \sqrt{\frac{eF}{MN\sigma^*(Bw)}}, \tag{4.59a}$$

Table 14. Parameters of formulas (4.59) [5].

M/m_a	A	B	$A_\perp$	$B_\perp$	$A_\parallel$	$B_\parallel$
0.1	0.514	4.32	0.343	4.60	0.195	3.43
0.2	0.627	3.21	0.306	3.24	0.185	2.41
0.5	0.859	2.12	0.249	1.83	0.158	1.38
0.8	1.042	1.70	0.224	1.43	0.148	1.02
1	1.147	1.55	0.212	1.25	0.146	0.91
1.5	1.366	1.35	0.196	1.04	0.146	0.77
2	1.548	1.26	0.187	0.93	0.148	0.73
3	1.851	1.17	0.182	0.84	0.152	0.69
4	2.109	1.12	0.182	0.83	0.153	0.65

$$D_\perp = A_\perp \frac{m_a}{M} \sqrt{\frac{eF}{M}} \frac{1}{[N\sigma^*(B_\perp w)]^{3/2}} , \tag{4.59b}$$

$$D_\parallel = A_\parallel \frac{m_a}{M} \sqrt{\frac{eF}{M}} \frac{1}{\left[N\sigma^*(B_\parallel w)\right]^{3/2}} . \tag{4.59c}$$

In the limits $n \to \infty$, $M/m_a \to 0$ formulas (4.59) are transformed into expressions (4.57), and at intermediate ratios of the ion and atom masses, the transport coefficients of ions result from solutions of the kinetic equation for ions. In Table 14 we give the results of this solution for $n = 8; 12$ [99], representing them in the form of relations (4.59) that allow us to extend these results to any large n.

5. Conclusions

In this paper we represent the asymptotic theory of resonant charge exchange that allows one to evaluate the cross section of this process by expanding it in a power series of a small parameter. This method enables one to both determine the cross section of the process and estimate its accuracy. The expansion method over a small parameter is a reliable approach because it gives the results in a simple form and evaluates its accuracy under specific conditions. We illustrate this by calculating transport coefficients in a gas in an external field if the cross section of ion – atom scattering depends weakly on the collision velocity. Then one can use a modified model of hard spheres wherein the final result is given in the simple form for the case when the cross section is independent of the collision velocity, but this cross section is related to a certain collision velocity which follows from the expansion of the result in a power series of a small parameter. This simplifies the result which is represented in the analytical form and the accuracy of such a simplification can be estimated. These methods may be extended to more complicate processes and problems.

REFERENCES

1. E. W. McDaniel and E. A. Mason, *The Mobility and Diffusion of Ions in Gases*, Wiley, New York (1973)

2. B. M. Smirnov, *Iony i Vozbuzhdennye Atomy v Plazme* (Ions and Excited Atoms in Plasma) [in Russian], Atomizdat, Moscow (1974)

3. H. W. Ellis, R. Y. Pai, E. W. McDaniel, et al., *Atom. Data Nucl. Data Tabl.*, **17**, 177 (1976)

4. H. W. Ellis, E. W. McDaniel, D. L. Albritton, et al., *Atom. Data Nucl. Data Tabl.*, **22**, 179 (1978)

5. A. A. Radzig and B. M. Smirnov, in: *Khimiya Plazmy* (Chemistry of Plasma) [in Russian], (edited by B. M. Smirnov), Vol. 11, Energoatomizdat, Moscow (1984), p. 170

6. H. W. Ellis, M. G. Thackston, E. W. McDaniel, and E. A. Mason, *Atom. Data Nucl. Data Tabl.*, **31**, 113 (1984)

7. E. A. Mason and E. W. McDaniel, *Transport Properties of Ions in Gases*, Wiley, New York (1988)

8. L. A. Viehland and E. A. Mason, *Atom. Data Nucl. Data Tabl.*, **60**, 37 (1995)

9. H. S. W. Massey and R. A. Smith, *Proc. R. Soc. London Ser. A*, **142**, 142 (1933)

10. H. S. W. Massey and C. B. O. Mohr, *Proc. R. Soc. London Ser. A*, **144**, 88 (1934)

11. L. A. Sena, *Zh. Eksp. Theor. Fiz.*, **9**, 1320 (1939)

12. L. A. Sena, *Zh. Eksp. Theor. Fiz.*, **16**, 734 (1946)

13. L. A. Sena, *Stolknoveniya Elektronov i Ionov v Atomakh* (Collisions of Electrons and Ions with Atoms) [in Russian], Gostekhizdat, Leningrad (1948)

14. Yu. N. Demkov, *Uch. Zapiski Leningr. Gos. Univ. Ser. Fiz. Nauk* [in Russian], No. 146, 74 (1952)

15. O. B. Firsov, *Zh. Eksp. Theor. Fiz.*, **21**, 1001 (1951)

16. B. M. Smirnov, *Zh. Eksp. Teor. Fiz.*, **46**, 1017 (1964). English translation: *Sov. Phys. JETP*, **19**, 692 (1964)

17. B. M. Smirnov, *Zh. Eksp. Teor. Fiz.*, **47**, 518 (1964). English translation: *Sov. Phys. JETP*, **20**, 345 (1965)

18. B. M. Smirnov, *Asimptoticheskie Metody v Teorii Atomnykh Stolknovenii* (Asymptotic Methods in Theory of Atomic Collisions) [in Russian], Atomizdat, Moscow (1973)

19. C. Herring, *Rev. Mod. Phys.*, **34**, 631 (1962)

20. L. D. Landau and E. M. Lifshitz, *Kvantovaya Mekhanika: Nerelyativistskaya Teoriya* (Quantum Mechanics: Non-Relativistic Theory) [in Russian], Nauka, Moscow (1974). English translation: Pergamon Press, Oxford (1977)

21. B. M. Smirnov, *Teplofiz. Vys. Temp.*, **4**, 429 (1966)

22. E. L. Duman and B. M. Smirnov, *Zh. Tekh. Fiz.*, **40**, 91 (1970). English translation: *Sov. Phys. Tech. Phys.*, **15**, 61 (1970)

23. E. E. Nikitin and B. M. Smirnov, *Usp. Fiz. Nauk*, **124**, 201 (1978). English translation: *Sov. Phys. Usp.*, **21**, 95 (1978)

24. G. Racah, *Phys. Rev.*, **61**, 186 (1942); **62**, 438 (1942)

25. E. U. Condon and G. H. Shortley, *The Theory of Atomic Spectra*, Univ. Press, Cambridge (1951)

26. I. I. Sobelman, *Atomic Spectra and Radiative Transitions*, Springer-Verlag, Berlin (1979)

27. D. A. Varshalovich, A. N. Moskalev, and V. K. Khersonskii, *Kvantovaya Teoriya Uglovogo Momenta* (Quantum Theory of Angular Momentum) [in Russian], Nauka, Leningrad (1975). English translation: World Scientific Publ., Singapore (1988)

28. B. M. Smirnov, *Phys. Scripta*, **61**, 595 (2000)

29. B. M. Smirnov, *Zh. Eksp. Teor. Fiz.*, **119**, 1099 (2001). English translation: *JETP*, **92**, 951 (2001)

30. A. I. Evseev, A. A. Radzig, and B. M. Smirnov, *Opt. Spektrosk.*, **44**, 833 (1978). English translation: *Opt. Spectrosc.*, **44**, 495 (1978)

31. A. A. Radzig and B. M. Smirnov, *Spravochnik po Atomnoi i Molekulyarnoi Fizike* (Reference Data on Atoms, Molecules, and Ions) [in Russian], Atomizdat, Moscow (1980). English translation: Springer-Verlag, Berlin (1985)

32. E. Clementi and C. Roetti, *Atom. Data Nucl. Data Tabl.*, **14**, 177 (1974)

33. A. D. McLean and R. S. McLean, *Atom. Data Nucl. Data Tabl.*, **26**, 197 (1981)

34. R. S. Mulliken, *Rev. Mod. Phys.*, **2**, 60 (1930)

35. F. Hund, *Z. Phys.*, **36**, 637 (1936)

36. E. E. Nikitin, *Opt. Spektrosk.*, **22**, 379 (1966)

37. E. E. Nikitin and S. Ya. Umanskii, *Neadiabaticheskie Perekhody pri Medlennykh Atomnykh Stolknoveniyakh* (Theory of Slow Atomic Collisions) [in Russian], Atomizdat, Moscow (1979). English translation: Springer-Verlag, Berlin (1984)

38. E. E. Nikitin and B. M. Smirnov, *Atomno-molekulyarnye Protsessy* (Atomic and Molecular Processes) [in Russian], Nauka, Moscow (1988)

39. B. M. Smirnov, *Usp. Fiz. Nauk*, **171**, 233 (2001). English translation: *Phys. Usp.*, **44**, 221 (2001)

40. H. B. Gilbody and J. B. Hasted, *Proc. R. Soc. London Ser. A*, **238**, 334 (1956)

41. Y. Kaneko, N. Kobayashi, and I. Kanomata, *J. Phys. Soc. Jpn*, **27**, 992 (1960)

42. A. Galli, A. Giardani-Guidoni, and G. G. Volpi, *Nuovo Cimento*, **26**, 845 (1962)

43. R. M. Kushnir, B. M. Paluch, and L. A. Sena, *Izv. Akad. Nauk SSSR, Ser. Fiz.*, **23**, 1007 (1959)

44. E. Gustafsson and E. Lindholm, *Ark. Fys.*, **18**, 219 (1960)

45. L. Kevan and D. L. Smith, *J. Am. Chem. Soc.*, **93**, 2113 (1971)

46. S. N. Grosh and W. F. Sheridan, *J. Chem. Phys.*, **26**, 480 (1957); *Indian. J. Phys.*, **31**, 337 (1957)

47. B. I. Kikiani, Z. E. Saliya, and I. G. Bagdasarova, *Zh. Tekh. Fiz.*, **45**, 586 (1975)

48. J. F. Williams, *Can. J. Phys.*, **46**, 2339 (1968)

49. E. L. Duman and B. M. Smirnov, *Teplofiz. Vys. Temp.*, **12**, 502 (1974)

50. E. L. Duman and B. M. Smirnov, *Teplofiz. Vys. Temp.*, **17**, 1328 (1979)

51. E. L. Duman, et al., "Charge Exchange Processes", Preprint of I. V. Kurchatov Institute of Atomic Energy, Moscow, No. 3532/12 (1982)

52. S. Sakabe and Y. Izawa, *Atom. Data Nucl. Data Tabl.*, **49**, 257 (1991)

53. S. Sakabe and Y. Izawa, *Phys. Rev. A*, **45**, 2086 (1992)

54. F. B. M. Copeland and D. S. F. Crothers, *Atom. Data Nucl. Data Tabl.*, **65**, 273 (1997)

55. L. D. Landau and E. M. Lifshitz, *Mekhanika* (Mechanics) [in Russian], Nauka, Moscow (1979). English translation: Pergamon Press, Oxford (1976)

56. B. M. Smirnov, *Dokl. Akad. Nauk SSSR*, **157**, 325 (1964)

57. E. M. Lifshitz and L. P. Pitaevskii, *Fizicheskaya Kinetika* (Physical Kinetics) [in Russian], Nauka, Moscow (1979). English translation: Pergamon Press, Oxford (1981)

58. T. Kihara, *Rev. Mod. Phys.*, **25**, 844 (1953)

59. B. M. Smirnov, *Fizika Slaboionizovannogo Gaza: v Zadachakh s Resheniyami* (Physics of Weakly Ionized Gases: Problems and Solutions) [in Russian], Nauka, Moscow (1972). English translation: Mir Publ., Moscow (1981)

60. B. M. Smirnov, *Dokl. Akad. Nauk SSSR*, **168**, 322 (1966)

61. G. H. Wannier, *Bell Syst. Tech. J.*, **32**, 170 (1953)

62. S. Chapman and T. G. Cowling, *The Mathematical Theory of Non-uniform Gases*, 2nd ed., Univ. Press, Cambridge (1952)

63. J. H. Ferziger and H. G. Kaper, *Mathematical Theory of Transport Processes in Gases*, North-Holland Publ. Co., Amsterdam (1972)

64. A. Dalgarno, M. R. C. McDowell, and A. Williams, *Philos. Trans. R. Soc. London Ser. A*, **250**, 411 (1958)

65. A. Dalgarno, *Philos. Trans. R. Soc. London Ser. A*, **250**, 428 (1958)

66. A. M. Tyndall, *The Mobility of Positive Ions in Gases*, The Univ. Press, Cambridge (1938)

67. R. W. Crompton and M. T. Elford, *Proc. R. Soc. London Ser. A*, **74**, 497 (1964)

68. T. Holstein, *J. Phys. Chem.*, **56**, 832 (1952)

69. B. M. Smirnov, *Atomnye stolknoveniya i Elementarnye Protsessy v Plazme* (Atomic Collisions and Elementary Processes in Plasma) [in Russian], Atomizdat, Moscow (1968)

70. Yu. P. Mordvinov and B. M. Smirnov, *Zh. Eksp. Teor. Fiz.*, **48**, 133 (1965). English translation: *Sov. Phys. JETP*, **21**, 98 (1965)

71. L. M. Chanin and M. A. Biondi, *Phys. Rev.*, **106**, 473 (1957)

72. O. J. Orient, *Can. J. Phys.*, **45**, 3915 (1967)

73. P. L. Patterson, *Phys. Rev. A*, **2**, 1154 (1970)

74. G. E. Courville and M. A. Biondi, *J. Chem. Phys.*, **37**, 616 (1962)

75. R. A. Gerber and M. A. Gusinow, *Phys. Rev. A*, **4**, 2027 (1971)

76. L. M. Chanin and R. D. Steen, *Phys. Rev.*, **132**, 2554 (1963)

77. Y. Lee and B. H. Mahan, *J. Chem. Phys.*, **43**, 2016 (1965)

78. G. Musa, A. Batlog, L. Nastase, et al., in: *Proc. of the 9th Intern. Conf. on Phenomena in Ionized Gases, Bucharest, 1969* (edited by G. Musa), Editura Academiei Republicii Socialiste România, Bucharest (1969), p. 10

79. A. Popesku and N. D. Niculesku, in: *Proc. of the 9th Intern. Conf. on Phenomena in Ionized Gases, Bucharest, 1969* (edited by G. Musa), Editura Academiei Republicii Socialiste România, Bucharest (1969), p. 8

80. T. M. Kojima, N. Saito, N. Kobayashi, and Y. Kaneko, *J. Phys. Soc. Jpn*, **56**, 6 (1992)

81. N. Saito, T. M. Kojima, N. Kobayashi, and Y. Kaneko, *J. Chem. Phys.*, **100**, 5726 (1994)

82. H. Helm, *J. Phys. B: At. Mol. Phys.*, **9**, 1171 (1976)

83. P. H. Larsen and M. T. Elford, *J. Phys. B: At. Mol. Phys.*, **19**, 449 (1986)

84. E. A. Mason and H. W. Schamp, *Ann. Phys.* (Leipzig), **4**, 233 (1958)

85. L. A. Viehland, M. M. Harrington, and E. A. Mason, *Chem. Phys.*, **17**, 433 (1976)

86. L. A. Viehland and S. L. Lin, *Chem. Phys.*, **43**, 135 (1979)

87. D. R. James, E. Graham, G. M. Thomson, and E. W. McDaniel, *J. Chem. Phys.*, **58**, 3652 (1973)

88. I. R. Gatland, M. G. Thackston, W. M. Pope, et al., *J. Chem. Phys.*, **68**, 2775 (1978)

89. V. I. Perel, *Zh. Eksp. Teor. Fiz.*, **32**, 526 (1957). English translation: *Sov. Phys. JETP*, **5**, 436 (1957)

90. B. M. Smirnov, *Dokl. Akad. Nauk SSSR*, **181**, 61 (1968)

91. R. J. Munson and A. M. Tyndal, *Proc. R. Soc. London Ser. A*, **177**, 187 (1940)

92. R. N. Varney, *Phys. Rev.*, **88**, 362 (1952)

93. E. C. Beaty, *Phys. Rev.*, **104**, 17 (1956)

94. A. B. Rakshit and P. Warneck, *J. Chem. Phys.*, **73**, 2673 (1980)

95. E. C. Beaty, in: *Proc. of the 5th Intern. Conf. on Ionization Phenomena in Gases, 28 Aug. – 1 Sept., 1961, Munich, Germany*, Vol. 1, North-Holland, Amsterdam (1962), p. 183

96. A. Einstein, *Ann. Phys.* (Leipzig), **17**, 549 (1905)

97. A. Einstein, *Ann. Phys.* (Leipzig), **19**, 371 (1906)

98. M. A. Leontovich, *Vvedenie v Termodinamiku. Statisticheskaya Fizika* (Introduction to Thermodynamics. Statistical Physics) [in Russian], Nauka, Moscow (1983)

99. H. R. Skullerud, *J. Phys. B: At. Mol. Phys.*, **9**, 535 (1976)

100. L. A. Palkina, B. M. Smirnov, and M. I. Chibisov, *Zh. Eksp. Theor. Fiz.*, **56**, 340 (1969). English translation: *Sov. Phys. JETP*, **29**, 187 (1969)